ROADMAP TO
GREENER COMPUTING

ROADMAP TO
GREENER COMPUTING

RAOUL-ABELIN CHOUMIN NGUEMALEU
LIONEL MONTHEU

Engineers-Pool
Hannover, Germany

CRC Press
Taylor & Francis Group
Boca Raton   London   New York

CRC Press is an imprint of the
Taylor & Francis Group, an **informa** business
A CHAPMAN & HALL BOOK

CRC Press
Taylor & Francis Group
6000 Broken Sound Parkway NW, Suite 300
Boca Raton, FL 33487-2742

Printed on acid-free paper
Version Date: 20140418

International Standard Book Number-13: 978-1-4665-0684-8 (Paperback)

Library of Congress Cataloging-in-Publication Data

Nguemaleu, Raoul-Abelin Choumin.
    Roadmap to greener computing / authors, Raoul-Abelin Choumin Nguemaleu and Lionel Montheu.
        pages cm
    Summary: "A concise and accessible introduction to green computing and green IT, this book addresses how computer science and the computer infrastructure affect the environment and presents the main challenges in making computing more environmentally friendly. The authors review the methodologies, designs, frameworks, and software development tools that can be used in computer science to reduce energy consumption and still compute efficiently. They also focus on Computer Aided Design (CAD) and describe what design engineers and CAD software applications can do to support new streamlined business directions and improve the environment"-- Provided by publisher.
    Includes bibliographical references and index.
    ISBN 978-1-4665-0684-8 (paperback)
    1. Data processing service centers--Energy conservation. 2. Computer systems--Energy conservation. 3. Information technology--Environmental aspects. 4. Electronic digital computers--Power supply. 5. Green technology. I. Montheu, Lionel. II. Title.

TJ163.5.O35N58 2014
004.028'6--dc23

2013045323

Visit the Taylor & Francis Web site at
http://www.taylorandfrancis.com

and the CRC Press Web site at
http://www.crcpress.com

Contents

Preface

THIS PROJECT BEGAN IN 2009, after an international engineering conference at Stanford University, California. On the flight back home to
Germany, we were reflecting on what lessons we had learned and what we
could take away from this inspiring experience. We asked ourselves what
we as an engineer and a computer scientist could do to positively impact
our community and the world.

Today, computers (our domains of expertise are software specialist, and
hardware specialist/ product lifecycle management specialist, respectively)
belong to the world heritage and are used by the most influential CEOs on
Wall Street as well as the poor orphan child in India or in Central Africa.

Originating from sub-Saharan Africa, we experienced firsthand the
negative impacts of global warming on agriculture, health, and the environment during childhood such as frequent floods and empty granaries
in the north region of Cameroon. Thus, to us, this book is a contribution
toward fighting the plague that afflicts our environment and humanity.
This book is a concise and simplified introduction to greener computing,
which can be easily accessed. It addresses how computer science, computer infrastructure, and the computer as product affect the environment,
and presents the main challenges involved in making computing more
environmentally friendly.

When we started writing the book, we were thinking of a one-theme
book with different related chapters to avoid repetition. After some discussions, we finally decided that the book should be divided into independent
chapters, such that each author should, beside his expertise, have a free
hand in introducing more general information in the project because it is
not only meant for specialists.

Before we started writing, we made an online survey of the question,
"What is a difficult book for you?" Based on the results of this survey,
we pointed out that people have encountered chapters or books that they

just cannot get into. There are lots of reasons for this: sometimes they are required to read about a topic that is just plain boring; sometimes they try to read material that is written way above their current intellectual level; and sometimes they find that the writer is not good at clearly explaining things. Our intention is that this book should be readable for a large audience, for kids in schools, teenagers in high school, college and university students, teachers, professors, homemakers, football players, computer specialists, taxi drivers, etc.

Lionel, who is a software engineer with wide field experience, focused on the computer infrastructure and software while Raoul, who has more than 12 years of field experience in software deployment and product lifecycle management, handled the part on the computer as a tool and as a product. As such, the book is divided into six independently written chapters to provide the reader with profound insights on each topic addressed.

CHAPTER 1: ON THE WAY TO ECO-FRIENDLY COMPUTER INFRASTRUCTURE

Here, the reader is informed about the impact of the computer infrastructure lifecycle on the environment and learns some solutions for greener IT. Ecological friendliness is gaining more and more importance nowadays. As the environment and natural milieus are negatively affected by human actions, science and engineering are increasingly conceiving and developing ideas that can make the world greener. The goal of this chapter is to present how computer science and more especially computer infrastructure affect the environment, what major challenges exist in this regard, and how they can be overcome or prevented. After a short introduction in Section 1.1, the problem of toxins in today's computers, their negative health and environmental effects, as well as some options for the manufacturing of ecologically (green) friendly computers are presented in Section 1.2. In Section 1.3, we analyze how making the right choices, while considering several criteria, can help in the purchase of green IT. Section 1.4 addresses power consumption and problems related to cooling when using computers in private or corporate settings. Section 1.5 covers challenges regarding the disposal of old computers or their components, and finally, in Section 1.6, some examples of how Information and Technology (IT) can positively influence the environment are described.

CHAPTER 2: GREEN SOFTWARE SCIENCE

Here, methodologies, designs, frameworks, and software development tools that can be used to compute energy efficiently are presented. According to the Green World Network, two of the world's biggest problems today are environmental damage caused by enormous greenhouse gas emissions and energy shortage, against the backdrop of the fact that the world's energy reserves are not unlimited. The Green World Network estimates that the Earth's supply of natural resources will be able to sustain only 2 billion humans for 100 years from 2000.* Thus, a reduction in energy consumption in all activities will greatly defuse these problems. In this work, methodologies, designs, frameworks, and software development tools that can be used in computer science to reduce energy consumption or compute energy efficiently are analyzed.

CHAPTER 3: COMPUTER AIDED SUSTAINABLE DESIGN

Designer and CAD applications that can reduce damage to products and the environment are discussed in this chapter.

Producing and consuming more and at the lowest possible price have characterized our society for several decades. This tendency has led to the overexploitation of natural resources, the intensification of air and water pollution, the disappearing of plant and animal species, and a dramatic increase in waste. In order to break this chain, urgent action must be undertaken such that more can be produced with less. Several corporations have embraced this new "produce more with less" approach and have started including sustainable development in their business strategies.

What can design engineers and CAD applications do to support the new "produce more with less" business approach and conserve the environment? In this chapter we analyze different approaches and propose answers to this question.

General information on design and the environment, access to literature, field experience, and concrete day-to-day business scenarios are discussed with the aim of getting a clear balance between theory and practice.

We also explain the Eco-Design concept and the impact of CAD on the environment as well as potential areas for improvement in the future.

---

* Reproduced by permission of the Australian Broadcasting Corporation (c) (2000) ABC. All rights reserved. http://www.abc.net.au/civics/environment/02_p1problems.htm (accessed January 21, 2013).

CHAPTER 4: COMPUTING NOISE POLLUTION

Here the reader is given an introduction to noise pollution. Together with light and heat, noise pollution is a form of energy pollution. Its contaminants are not physical; it is therefore sometimes difficult to be fully aware of this "silent" destructor, which has become a plague to an increasing proportion of the population. Noise generated by PC fans or air cooling systems has always been an issue for people working intensively with computers (stress engineer, computer gamers, data center administrator, etc.). This chapter discusses noise pollution caused by computers and computer users, its effects on their daily lives, and solutions that can be employed to counteract it.

The first section of this chapter focuses on the definition of sound in general, its various sources, and mechanisms for its perception by humans, that is, via sound sensors in the ear and the brain. In addition, the effects of sound (as noise) on the mind and body are discussed. The second section examines computer noise and standards, techniques, tips, and tricks that can be employed to counteract it. Some practical examples are also included in this section.

CHAPTER 5: END-OF-LIFE OPPORTUNITIES FOR COMPUTER AND COMPUTER PARTS

In this chapter, we discuss basic ways of handling a computer when it begins to become obsolete for the end user. We propose a flow diagram representing the different given options, when a computer no longer fulfills its purpose.

It goes without saying that having a well-established computer or series of computers on the market is the main reason why computer companies develop and improve their products. Producers therefore tend to ensure the supply of successful products for as long as possible. The more computers are produced, the more effort must be invested to take them out of life in an environmentally friendly way. At the end of their lives, many computers are deposited in landfills while many more are upgraded or recycled as markets for used computers and electronic devices expand. Yet many are retained, despite the disadvantages of storage.

CHAPTER 6: CLOUD COMPUTING

In this chapter, the reader is introduced to cloud computing by way of definition and an overview of its green capabilities. Nowadays, it is quite hard not to have heard of cloud computing, even as a layperson. It is even likely that you use cloud-computing services without knowing it.

The overall demand for new, state-of-the-art IT services has increased and in order to continuously offer innovative services and at the same time fulfill financial requirements, it is relevant to ask whether we should rely on cloud computing or standard IT infrastructure. We first describe how computing was carried out before the cloud came and then go ahead to explain what cloud computing means. We describe the characteristics and specific capabilities of clouds and take a deeper look into existing services and applicable deployment models. We also discuss several use cases of cloud computing, its benefits, and potential risks. Finally, we look into if cloud computing can be green.

Acknowledgments

W E INVESTED A LOT of effort in this project. However, it would not have been possible without the kind support and help of many individuals, especially our publisher team (Randi Cohen and Amber Donley) and the members of the "Engineers-Pool e.V.," a non-profit professional organization of engineers based in Germany.

We also wish to express our profound gratitude to our proofreader, Dr. Wendy Awa, for her helpful discussions, comments, and corrections, and useful suggestions, which have enabled us to complete the project on time. We would also like to express our gratitude to our families for their kind cooperation and encouragement throughout this project.

Note: The tests and experiments mentioned in this book were conducted by a team of professionals. If you wish to reproduce any of these tests and experiments by yourself, please seek professional assistance.

About the Authors

 **Raoul-Abelin Choumin Nguemaleu** was born and raised in Cameroon (Central Africa). After earning his Baccalauréat (high school diploma) in Cameroon, he went to Germany for university studies. Raoul is graduated in mechanical engineering from Hamburg University of Applied Sciences in Germany.

He started his professional career as a CAD/PLM consultant working for the world's two leading aircraft manufacturers. His assignment focused on the development of appropriate, accurate, and valuable CAD/PLM training, processes, and methods. He has also worked in many companies as a consultant for software testing and implementation. In this function, he was a member of the CAD/PDM/Configuration management team for many projects in Germany, UK, France, and the United States.

For the last five years, he has worked as a global service consultant at one of the largest software companies in the world and the leading provider of product development solutions. Today, he works as CAE Coordinator for a global leader in aerospace, defense, and related services. He is a member of several environmental organizations.

Lionel Montheu was born and raised in Cameroon (Central Africa). He visited the school there and decided to come to Germany after earning his Baccalauréat (high school diploma). Lionel is graduated in computer science from University of Ulm (South Germany).

While working on his diploma thesis, he was involved with the optimization of monitoring software in an embedded system used in satellite ground stations. He started his professional career as an IT solution designer, and developed software solutions and tools used by different departments in a universal and international trading bank in Germany. Due to his position, he was involved in different phases of the software development process (requirements engineering, conception, design, implementation, test, and maintenance).

Today he works as a software developer for a company that has proprietary software specialized in business service monitoring. He is also a member of several environmental organizations.

CHAPTER 1

On the Way to Eco-Friendly Computer Infrastructure

The Impact of Computer Infrastructure Lifecycle on the Environment and Solutions for Greener IT

DEFINITIONS AND ABBREVIATIONS

ATA: Advanced Technology Attachment
BTU: British thermal unit (traditional unit for measuring energy)
CD: Compact disc
CO_2: Carbon dioxide
CPU: Central processing unit
DNA: Deoxyribonucleic acid
DVD: Digital Video Disc/Digital Versatile Disc
DX: Direct eXpansion
E-Waste: Electronic waste
F: Fahrenheit (unit of temperature)
HVAC: Heating, Ventilating, and Air Conditioning
IP: Internet Protocol

IT: Information and Technology

kWh: Kilowatt per hour

LCA: Lifecycle Analysis/Lifecycle Assessment

LCD: Liquid Crystal Display

PC: Personal computer

PVC: Polyvinylchloride

QoS: Quality of Service

RAM: Random Access Memory

RoHS: Restriction of Hazardous Substances

SAN: Storage Area Network

SCSI: Small Computer System Interface

TCO Certification: Eco label granted by the TCO (Swedish Confederation of Professional Employees) Development

USB: Universal Serial Bus

W: Watt (derived unit of power)

WEEE: Waste Electrical and Electronic Equipment

1.1 INTRODUCTION

Today, for many people, it is hard to imagine a world without computers and the Internet. Current advancements in technology allow us to achieve most things from a computer. For instance, very few people still communicate using postal means, as it is now widely common to stay in touch via e-mail. It is even possible to make phone calls, watch videos, discuss in conference rooms, and play games on a computer via the Internet. Such undertakings are possible thanks to services like voice over internet protocol (VoIP: internet phone calls), streaming (watching movies and listening to music online), social networking, and much more. The availability of these services relies heavily on strong IT infrastructure.

Although we benefit from the possibilities that computers offer, we are also negatively affected by the impact of IT infrastructure on our environment. The lifecycle of IT infrastructure consists of manufacturing, purchasing, use, and disposal/recycling. In this work, we will examine why and how these lifecycle phases affect our environment. We will also propose concrete solutions that can help make the lifecycle of IT infrastructure more environmentally friendly.

1.2 MANUFACTURING COMPUTER COMPONENTS

The initial phase in the lifecycle of computer components is their manufacture. As early as this point, the choice of adequate materials and chemicals can impact the environment positively or negatively.

1.2.1 Current Situation

Figure 1.1 is an illustration of a desktop computer with its basic components such as a keyboard, monitor, mouse, and the tower hosting the central processing unit (CPU).

The numbered elements show where various toxins can be found in personal desktop computers [1] or in computers used in corporations. Next we describe the negative effects of computer infrastructure on human health.

1.2.1.1 Lead

Many older TV and computer monitors contain lead, which is also used in soldering of circuit boards. Exposure to lead [2] can be through the inhalation of lead dust or fumes, ingestion of lead-contaminated food or water, or through contaminated hands, cigarettes, or clothing. When lead enters the respiratory and digestive systems, it passes into the blood and is distributed throughout the body. Following contamination, more than 90% of the total body burden of lead accumulates or is stored in bone. Lead in bones may later be released into the blood system, re-exposing organs to damage long after the initial exposure. Lead can affect all organ and body

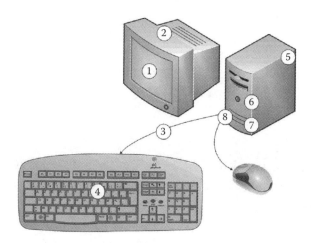

FIGURE 1.1 A basic computer system.

functions in varying degrees. Particularly susceptible are neurological, gastrointestinal, reproductive, and renal organ systems.

1.2.1.2 Arsenic
Arsenic can be found in old cathode ray tubes and many of its compounds are potentially highly poisonous. In the European Union (EU) elemental arsenic and arsenic compounds are classified as "toxic" and "dangerous" to the environment. Arsenic causes severe metabolic interferences, which can lead to death from multi-organ system failure.

1.2.1.3 Polybrominated Flame Retardants
Brominated flame retardants are used in plastic casings and are released when electronics are dumped or incinerated. They are likely a cause of endocrine dysfunction, for example, reduction in the levels of the hormone thyroxin in exposed humans/animals and potential endangerment of fetal development in pregnant women.

1.2.1.4 Selenium
Selenium is used as a power supply rectifier in circuit boards. This element can cause selenium poisoning in case of excessive exposure. A selenium dose as small as 5 mg per day is potentially lethal for humans.

1.2.1.5 Cadmium
Infrared detectors, semiconductors, older types of cathode ray tubes, and some plastics contain cadmium. Cadmium accumulation [2] in the body can cause kidney and bone damage, which can lead to cancer.

1.2.1.6 Chromium
Chromium is used to protect untreated and galvanized steel plates from corrosion and in hardening steel housing. It can cause DNA damage and asthmatic bronchitis.

1.2.1.7 Mercury
Mercury can be found in light bulbs, flat panel displays, LCD screens, switches, and printed wiring boards. High levels of exposure contribute to brain and kidney damage, cause harm to developing fetuses, and can be passed to infants through breast milk. Mercury contamination can also occur through the consumption of fish. Exposure through ingestion or inhalation can cause central nervous system and kidney damage.

Component or material type	Highest value[a]	Number of samples with bromine (total number of samples analysed) concentration range (%)[b]					
		Dell	Sony	Apple	Acer	HP	Toshiba
Circuit boards	Acer 8.2%	15 (15) 1.8–6.2%	1 (10) 6.0%	22 (24) 0.15–7.4%	20 (24) 2.0–8.2%	21 (24) 1.9–7.5%	2 (5) 6.6–7.7%
Ribbon cables[c]	HP 7.0%	9 (12) 0.58–2.7%	– (8) –	8 (12) 1.1–6.8%	11 (15) 0.2–3.0%	7 (16) 2.2–7.0%	3 (4) 1.7–2.2%
Insulation sheets	Dell 5.8%	6 (6) 1.4–5.8%	(4) –	4 (8) 2.8–2.9%	1 (8) 1.9%	– (8) –	– (2) –
Internal plastic pin connectors	Dell/Acer 10%	5 (12) 3.7–10%	3 (8) 9.3–9.4%	5 (16) 0.2–7.4%	8 (16) 4.5–10%	5 (16) 0.26–7.0%	2 (4) 4.8–8.7%
Fans	Apple 6.6%	3 (3) 5.9–6.4%	2 (2) 4.6–6.1%	4 (4) 5.3–6.6%	4 (4) 4.5–6.4%	4 (4) 4.4–5.3%	1 (1) 5.9%
Socket for external battery	Apple 9.9%	3 (3) 7.9–8.5%	– (2) –	4 (4) 6.9–9.9%	4 (4) 8.4–8.8%	4 (4) 8.1–9.5%	1 (1) 9.2%
Main chip surface	HP 0.42%	3 (3) 0.30–0.38%	1 (2) 0.38%	4 (4) 0.30–0.38%	4 (4) 0.28–0.36%	4 (4) 0.29–0.42%	1 (1) 0.36%
External battery casing	Acer/HP 1.4%	1 (3) 0.64%	– (2) –	– (4) –	1 (4) 1.4%	1 (4) 1.4%	– (1) –
Power transformer casing	Sony 2.8%	1 (3) 0.30–0.35%	1 (2) 2.8%	– (4) –	1 (4) 0.18%	– (4) –	1 (1) 0.32%
Touch mouse pads	Apple 5.0%	1 (3) 2.2%	– (2) –	4 (4) 4.6–5.0%	– (4) –	– (4) –	1 (1) 0.54%
Mouse button	Dell 1.5%	1 (3) 1.5%	– (2) –	– (4) –	– (4) –	– (4) –	– (1) –
Plastic housing/casing	Apple 0.68%	– (15) –	– (8) –	1 (16) 0.68%	– (20) –	– (20) –	– (5) –

FIGURE 1.2 Summary of bromine content in samples of models from each brand, for key component groups analyzed. (a) Brand yields the highest surface bromine concentration for each component type; (b) all values above detection limit of 0.1%; (c) no surface bromine detected in sample 6 (ribbon cable connected to keyboard) for all laptops. (*Source*: Toxic Chemicals in Computers Reloaded. With the permission of Greenpeace.)

1.2.1.8 Plastics and Polyvinyl Chloride (PVC)

An average computer is made up of approximately 20% PVC. When PVC is burned, dioxin is formed. Combinations of plastics that are difficult to separate and recycle are used in printed circuit boards as connectors, plastic covers, or cables.

A Greenpeace study [3] shows bromine concentration in several laptop components of different manufacturers (Figure 1.2).

Computer components are often thrown away after a certain time period for different reasons such as malfunction, end of life due to material upgrade, etc. Throwing those components away produce electronic waste. In case of accidental breakdown of an electronic device, toxin contaminants can be released.

1.2.2 Some Alternatives

An alternative to harmful e-waste can be the reduction of toxins in the process of manufacturing components. This is referred to as green manufacturing. This approach is pursued by an EU decree, Restriction of Hazardous Substances Directive (RoHS), aimed at reducing the use of hazardous materials, which now pose a central concern for electrical and electronic manufacturers. This EU decree (in conjunction with the WEEE regulations) restricts the use of certain unsafe substances like lead and mercury in electrical and electronic equipment. This implies that the allowed amounts of these substances in computer components are reduced to substantially safer amounts. This program has been adopted globally under RoHS guidelines and sometimes under new program labels in various regions.

RoHS works in conjunction with the EU WEEE directive, supporting WEEE by reducing the amount of hazardous chemicals used in production processes. In turn, it reduces the risk of exposure for recycling staff as well as recycling costs. Manufacturers will need to ensure that their products, parts, and components comply with RoHS standards before they can be distributed and sold in the EU.

Another approach to this problem is using eco-friendly materials to make computers. Alternative materials should be renewable or produced using less energy.

- The use of bamboo to make computer casings and peripheral parts, as is the case with USB flash devices, is becoming more and more popular. Bamboo that has been cut only takes about five years to grow again. Bamboo furniture also looks good and is not heavy to carry.

- Instead of manufacturing computers using non-recyclable plastics, recyclable polycarbonate resins could preferably be used.

- As described previously, flame retardants are often the most toxic chemicals in traditional PCs. Alternatively, one could use flame retardant silicone compounds, which are readily available, to achieve the same function and they are completely non-toxic.

- A component break can happen during the recycling process, which could be harmful for workers if the component contained toxins. Reducing or (if possible) avoiding the use of toxins would make a great difference in such cases.

The main goal in this initial phase should be to *make the input green* while manufacturing computer systems. This involves the following options:

- Use of non-hazardous material

- Use of easily recyclable material

- Saving energy and resources while manufacturing

- Using recyclable material if possible

1.2.3 Lifecycle Analysis

It is important to keep the whole life cycle of a product in mind in order to have a better overview of its possible impact on the environment. This can be done by a Lifecycle Analysis (LCA).

LCA [4] is a technique for assessing potential impacts on the environment associated with a product as follows:

- Taking inventory of relevant inputs and outputs of a product system

- Evaluating the potential environmental impacts associated with inputs and outputs

- Interpreting results of inventory analysis and impact assessment phases in relation to the study objectives

Once the product is manufactured, it is released in the market and enters the next phase of its lifecycle.

1.3 GREEN PURCHASING IN IT

People mostly base their purchase decisions on several criteria including quality, price, and ease of use. Green purchasing is checking the effect a product has on the environment during its entire lifecycle before buying. If every customer (government, corporate, or private person) considers this aspect during each purchase decision, manufacturers will surely change their production philosophy to a greener one.

There is no general rule of thumb for green purchasing, which applies to every product. Nevertheless, focusing on certain criteria could make purchasing IT infrastructure (computers, peripherals, printers, copiers, related electronics, servers, etc.) environmentally friendlier.

FIGURE 1.3 The Energy Star logo. (*Source*: http://commons.wikimedia.org/wiki/File:Energy_Star_logo.svg. Public domain.)

- Energy efficiency: Power consumption of a computer usually generates more cost during its useful lifecycle than when the machine was purchased. Consumers may rely on international standards like Energy Star (Figure 1.3) (used for rating products based on energy efficiency) during the purchase of electronic components. This rating is calculated using several criteria, for example, the screensaver must be turned on if a computer has been in an idle state after 15 minutes. Since each corporation can compile its own rating and then inform the Environmental Protection Agency (which does not carry out any checks), customers can only hope for correctness and fairness of such ratings. Many European countries use different standards. For example, the Swedish Confederation of Professional Employees uses TCO Certified to label eco-friendly IT equipment. As stated on the TCO Development website, TCO Development works with many international committees and specialty networks, to ensure worldwide applicability for TCO Certified. The criteria in this standard are based on scientific studies and developed with focus on benefiting the user and the environment. TCO Certified is a third-party certification and works as follows:

  - A manufacturer sends a product to an independent laboratory.

  - The product is tested rigorously by the laboratory according to internationally accredited test methods.

  - TCO Development checks the results and eventually approves the use of the TCO Certified label.

- Nontoxic products: To avoid side effects resulting from exposure to computer components containing toxins, it would be better to use toxin-free hardware in the first place.

- Reparability: It is easier to replace single components than whole computer systems. Therefore, modular computer systems are preferable.

- Recyclability: Focus on systems that contain highly recyclable components/materials and can be easily disassembled and replaced.

- End-of-life program: Some manufacturers offer end-of-life programs that enable customers to return purchased products when they are no longer needed, such that manufacturers can recycle themselves.

- "Old" against "new" strategies: Sometimes people hesitate before buying new products (with enhanced functionality). The question that usually arises is what would happen to the old functional one? To attract such customers, several shops offer "old" against "new" programs, which give clients a price reduction when they return an old product before purchasing a new one. In doing so, the shops get new customers as well as promote themselves as discarding returned products in an eco-friendly way.

- Avoid green washing: Many companies claim to make green products and use ads and marketing campaigns to mislead consumers by overstating claims of their favorable environmental performance. Vague terms like "eco-friendly" or "environmentally preferable" are sometimes used to make products seem greener than they are. For example, a flat screen TV can be marketed as lead-free, which is actually true and sounds good because lead is toxic. However, the manufacturer does not mention that the flat screen contains mercury, another toxic chemical and the customer may decide to purchase the new TV without knowing about its toxicity.

LCA can also help in making green purchasing decisions. Customers aware of the product life would better recognize the difference between green products and those that negatively affect the environment. After a product has been purchased, it enters the next phase of its lifecycle.

1.4 USING COMPUTER SYSTEMS

The next phase in the lifecycle of computer systems is the operative phase. In this phase, computer components have been purchased by customers and are in use. Here, there are also several issues regarding the impact on the environment, the most often discussed being power consumption and cooling components/systems.

1.4.1 Power Consumption

Power consumption is one major issue, which is responsible for high electricity bills in households or corporations. This subsection will discuss how much power costs and why so much power is consumed in homes and corporate settings.

1.4.1.1 How Much Does Power Cost?

Power is measured in watts (W), a derived unit that measures the rate of energy conversion and is defined as 1 joule per second. Electricity is paid for by kilowatt hour (kWh). A simple example is used here to illustrate this. A 100 W light bulb uses 100 W of electricity in 1 hour. Therefore, 10 such bulbs are needed to achieve power consumption of 1 kW per hour. The price of power is not static but varies all over the world. Figure 1.4 shows average prices of electricity in different parts of the United States.

1.4.1.2 Composition of a Single Computer

Following are some basics on computer hardware. Figure 1.5 shows the main components of a personal computer (PC).

1. Motherboard: The motherboard performs fundamental function of a computer. It contains internal components (Central Processing Unit [CPU], memory, etc.) and manages the interface to external input/output devices.

2. Central Processing Unit (CPU): The CPU is the central component of a computer and allows data manipulation and processing.

3. Random Access Memory (RAM): RAM is a volatile memory that stores data processed by the CPU.

4. Hard drives: Hard drives are used to store user data and operating systems.

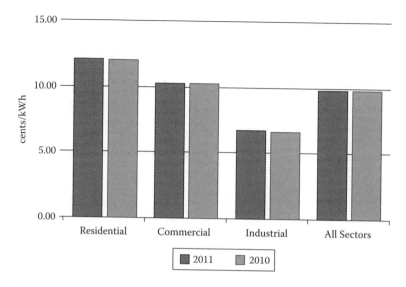

FIGURE 1.4 Average retail price of electricity to ultimate customers by end-use sector, year-to-date: through May 2011 and 2010. (*Source*: U.S. Energy Information Administration.)

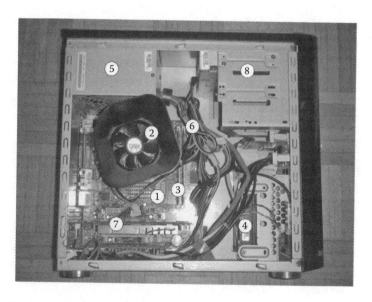

FIGURE 1.5 Internal constitution of a personal computer.

5. Power supply: Electrical energy is necessary for the functionality of internal components.

6. Power cables: Power cables are used to direct power from a power supply to a specific component.

7. Graphic card: A graphic card allows for the processing and display of computer data on a monitor.

8 CD/DVD drive: CD/DVD drives allow computers to read CDs and DVDs.

Without power, there is no life in a computer. Once power is turned on, electrical energy is transmitted from the power supply via cables to the motherboard, the CPU, the hard drive, CD drives, etc. The PC then boots and is ready for use.

The above demonstrates the use of power by a single PC. The more additional hardware that is needed, the more power that is required in order to fulfill additional requirements. This is true for both private and corporate computer infrastructure use.

1.4.1.3 Private Use

Nowadays, computers are used more often than they were 10 or 20 years ago. Thus, if you plan to buy a new computer, it is wise to think about aspects like power consumption, as this will affect electricity bills. Depending on how you plan to use the computer (Internet tasks, home office, gaming, etc.) you may need additional hardware like a webcam, a printer, or maybe an external hard drive and efficient processors.

Additional hardware usually means an additional component that will need power to run. Figure 1.6 shows a usual computer environment.

Table 1.1 shows the overall (estimated based on the power consumption of the embedded hardware components) power usage of a standard single PC with an integrated graphic chipset.

The estimated power consumption of a PC is shown in Table 1.2.

This demonstrates that having powerful graphics actually increases power consumption. Therefore, it is wise to pick the right components that correspond to your user profile and that would avoid energy losses and unnecessarily high power bills.

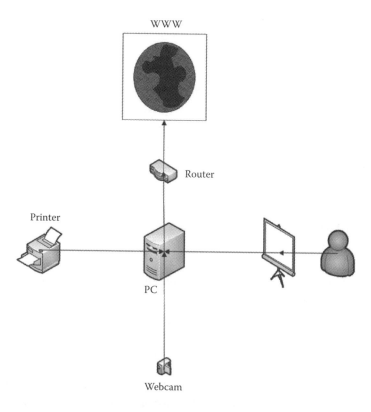

FIGURE 1.6   A single computer environment.

TABLE 1.1   Power Consumption of a Standard PC with an Integrated Graphic Chipset

Standard PC without Graphics Card	Power Usage (W)
Dual-core CPU	65
Motherboard with integrated graphics	20
2 Memory modules	6
2 Hard drives	20
Drive + burner	20
Total	**131**

1.4.1.4 Commercial/Corporate Use

In Section 1.4.1.3, we described power consumption in home use. Now we look at power consumption in the industrial setting.

Even in smaller enterprises, computers have become essential tools for performing daily tasks. The more computers and computer hardware

TABLE 1.2   Power Usage of a PC with a Standalone Graphic Card

Power PC with a Standalone Graphic Card	Power Usage (W)
Overclocked quad-core CPU	130
Motherboard with separate graphic card	60
4 Memory modules	12
4 Hard drives	40
Drive + burner	20
Total	**262**

are used, the more power is consumed. There are many reasons why IT infrastructure is needed. These include:

- Data redundancy. Imagine that you finished writing an important report and the day after you reach your jobsite to find out that the hard disk storing your data has crashed. This would be very annoying. To avoid such scenarios, important data can be stored redundantly (in data servers) to serve as backups (Figure 1.7).

- Server redundancy. A typical example for this use can be a simple email server. Emails are usually stored in several mail servers, so that if one of the servers is unavailable, the email can still be available on another server.

- Performance. Some applications are very CPU intensive and it can take a while before the results of a specific process are available. In time critical cases, when you do not want to wait hours for results, you can let the application run on several computers (grid computing, see Figure 1.8). Using an appropriate number of grid nodes will have a noticeable effect on the performance of the process.

FIGURE 1.7   Backup data on another server.

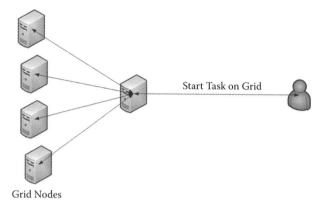

Start Task on Grid

Grid Nodes

FIGURE 1.8 Computing on several machines.

- Load balancing. When applications are deployed on dedicated servers, it is good to reduce the server and network load. This way, you avoid a scenario where all users are always routed to the same server when they start an application. Figure 1.9 shows an example of load balancing between several users and applications. This is common for published desktop applications.

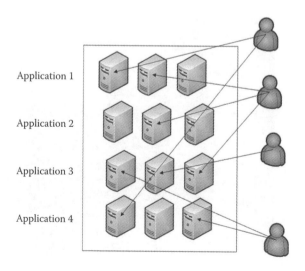

Application 1

Application 2

Application 3

Application 4

FIGURE 1.9 An example of load balancing.

1.4.1.5 Lower Power Consumption without Compromising Quality of Service (QoS)

In the industry, especially, power consumption is a huge problem and reducing power consumption does not only result in green benefits but also reduces costs. Because a good, efficient, and qualitative IT infrastructure comes with a price, our concern is to lower power consumption without compromising QoS. It is possible to optimize power consumption via the following means:

- Monitoring power usage. This should be the first place to start. Once you know what the current consumption level of a given hardware (monitor, printer, server, etc.) is, you can look for a suitable alternative and compare both. For private use, the tool Kill-A-Watt (Figure 1.10) is practical. You can plug it into the equipment to be monitored and then connect to a power source, such that the current consumption is displayed. Since Kill-A-Watt is a single monitoring tool, it is not applicable for corporate use. In such settings, software solutions are heavily relied upon. There are currently several tools for this purpose. Processing a web lookup via a search engine will guide you to some useful results.

1.4.1.5.1 Better Power Management   After using a computer at home, we usually turn it off. In some organizations, however, computers and monitors stay in the idle state even at night when they are not in use. If

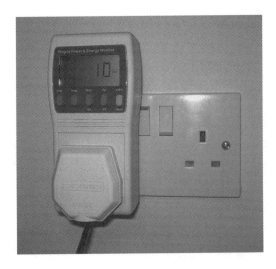

FIGURE 1.10   Kill-A-Watt. (*Source*: Public domain.)

employees get into the habit of turning off their monitors (and the computer as need be) when they leave for the day, this would prevent energy wastage. Sometimes it may be necessary to leave the computer turned on overnight. This is the case when software updates must be installed during a time frame when no users are connected to their devices. In order to ensure the smooth completion of such tasks, it may be helpful to inform users not to shut down their computers before leaving. Another alternative may be to let the computers check for updates when they are started and install them if necessary. In this case, if a user shuts down his or her device, despite a notification he or she will have to wait for the full update upon the next startup.

- Reduce standby power usage. The idea of standby power usage follows the One Watt Initiative, which is a proposal for energy saving by the International Energy Agency (IEA). Its goal is to reduce standby power usage in all appliances to 1 W. Many [5] domestic appliances and commercial equipment still consume electric power when they are switched off or not performing their primary function. Several initiatives to reduce standby losses have been introduced in different parts of the world.

- Storage area network vs. direct attached storage. It is also possible to optimize data storage. Some organizations use directly attached data storage. In this case, data are usually stored on file servers and each server has its own hard drive directly attached to it (e.g., via USB, fire wire, ATA, SCSI, etc.). Increasing the hard drive space mostly results

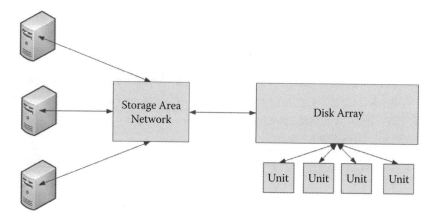

FIGURE 1.11 Architecture of a Storage Area Network (SAN).

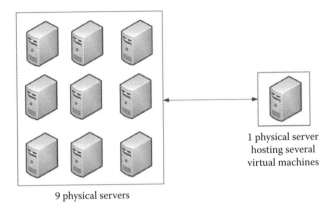

1 physical server
hosting several
virtual machines

9 physical servers

FIGURE 1.12   Virtualization.

in adding new hosts to the file servers. Another efficient way to save data would be the use of a Storage Area Network (SAN; Figure 1.11). Here you just have to add more disks (units) to the disk array to increase the hard drive space.

- Virtual machines instead of physical servers. As described in Section 1.4.1.4, there are several practical cases where having multiple servers is helpful in achieving desired tasks. However, physical servers are not always necessary. Since the problem of energy consumption grows with the number of machines used, the question arises, why not switch to virtual servers? Thanks to efficient virtualization software and advanced clustering technologies, it is possible to combine several machines into one single server, which itself can intelligently manage a specified number of virtual servers (Figure 1.12).

- Blade servers. The previous example can be optimized by the use of blade servers. An advantage of using a blade server is that it can be grouped in a rack. As such, it occupies less space and shares power usage (Figure 1.13).

- Automate night tasks. Some people leave their computers on the whole night in order to download a file or finish a task. The feature in a well-known operating system called "Schedule Task," which allows waking a computer from sleep mode (Figure 1.14) and performing specific tasks, can be used in such cases.

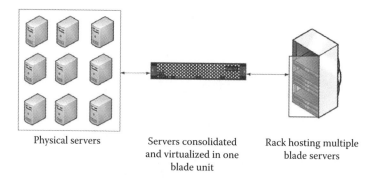

Physical servers   Servers consolidated and virtualized in one blade unit   Rack hosting multiple blade servers

FIGURE 1.13   Using blade servers in a rack.

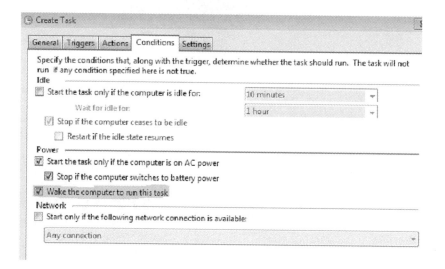

FIGURE 1.14   Waking computer from sleep mode using a scheduled task.

- CO_2 Saver. This is an application that runs in the background of Windows computers and adjusts power settings automatically. If a PC has not been used in a while, the CO_2 Saver will jump in and lower your setting's performance.

1.4.2 Cooling

Another major aspect with regard to IT components is cooling. As described previously, computer components need power to run. The use of power generates heat. Thus, the generated heat needs to be handled to prevent component damages resulting from overheating (overheated parts

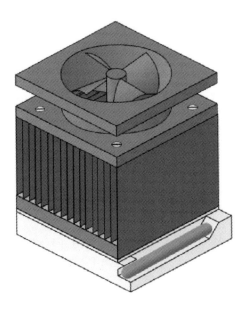

FIGURE 1.15 Cooling a single CPU. (*Source*: www.wpclipart.com. Public domain.)

fail early or may cause systems to freeze or crash). Figure 1.15 shows an example of a fan attached to a CPU.

1.4.2.1 Cooling Challenges at Data Centers
The problem of cooling is just as relevant in data centers as well. There is increased challenge in heat management and high cost of cooling in such settings, which come about for several reasons:

- More servers are in use in data centers for different purposes.
- Heat density is increased in server racks.
- There is irregular distribution of heat load in the data center.
- Cost of power is consequently high.

Every year, the information technology sector in the United States spends almost $7 billion on electricity costs, and much of that money goes to cooling CPUs in data centers [6].

TABLE 1.3   Where Power Is Used in Data Centers

Running servers	50%
UPS, transformers, and distributing losses	10%
Lighting	3%
Chilling water or compressing coolant to cool air	4%
Cooling servers	7%
Humidifying and dehumidifying	15%
Wasted by mixing in the room	7%
Design and system inefficiencies	4%

Energy can be used in data centers for the following purposes:

- run servers

- cool servers

- humidification and dehumidification

- light

- chilling water

However, other factors like mixing in server rooms, design inefficiencies, or system inefficiencies are also responsible for energy waste in data centers (Table 1.3).

Although overcooling the servers will also generate unnecessary extra costs, it is important to avoid under-cooling; otherwise, premature hardware failures will occur. It is appropriate and important to calculate how much cooling you need before buying new cooling equipment. The heat load of the infrastructure is not the only factor to be considered before decision-making. The following additional factors are also relevant [1]:

- Room size. For a server room having windows, the following formulas are useful to help determine what is best for your data center:
  - The south window's BTU can be calculated by multiplying the south facing window length (in meters) by the width (m) and by a factor of 870.

- The north window's BTU can be calculated by multiplying the north facing window length (in meters) by the width (m) and by a factor of 165.

- Then multiply the results by 1.5 if there are no blinds on the windows.

- People in the room. In a server room where people are permanently seated, the heat goes up to about 400 BTU per person. The total occupants' BTU can be found by multiplying the number of occupants by 400.

- Equipment. Most of the heat is generated from your equipment. You can find the equipment's power consumption in its original purchase brochure. Also, take into consideration all additional equipment that are or may eventually be stationed in the room. The equipment's BTU is then calculated by multiplying the total wattage for all equipment by a factor of 3.5.

- Lighting. The lighting's BTU is calculated by multiplying the total wattage for all lighting by the factor 4.25. The sum of the calculated values can then be used to get the amount of cooling needed and appropriate air conditioning can be purchased.

1.4.2.2 Reduce Cooling Costs in a Data Center

One can rely on natural conditions to lower the cost of cooling. For example, when surrounding temperatures and humidity are favorable, the use of air from outside for cooling is possible. In this case, an economizer system is needed. There are currently two types of economizers:

- Airside economizer. An airside economizer regulates the use of outside air to cool a data center. Thanks to sensors, the system finds out when the air outside is cold enough to be used in cooling the data center. When this is the case, the outside air is pulled in and distributed into cabinets via an existing air delivery system (Figure 1.16). Then there is no longer need for an air conditioning system. With this solution, it is important to be mindful of change in room humidity and also the fact that polluted air can be conducted into the data center. A properly operating economizer can cut [7] annual cooling costs by as much as 20%, depending on local climate and internal cooling load. Unfortunately, economizers are notoriously unreliable and break down frequently. Previous results from field

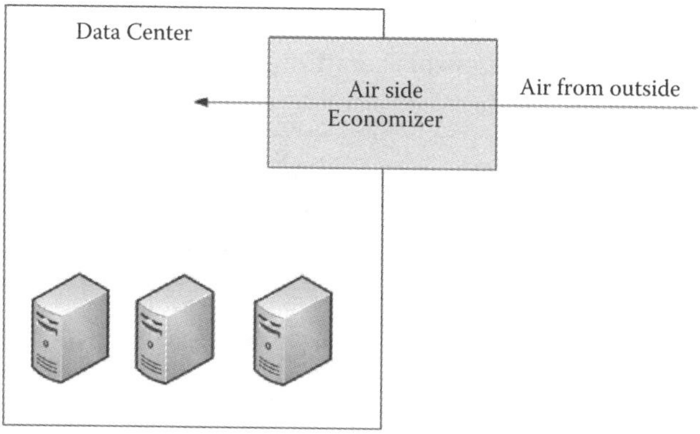

FIGURE 1.16 Where power is used in data centers.

studies suggest that about one-half of all newly installed economizers did not work properly. Reasons why both old and new economizers fail include:

- Jammed outside-air dumper

- Jammed, broken, or disconnected linkage

- Non-functional actuator

- Inaccurate air temperature sensors

- To make things worse, a malfunctioning economizer may become a major energy waster. A significant amount of energy may be used to heat or cool excess outdoor air, when an outside-air damper is stuck in an open position.

- Solution: Test economizers at least twice a year using the following procedure:

  - Check to see if the economizer section is running at minimum outdoor air settings when the system is mechanically cooling.

  - Cycle the minimum position potentiometer from zero to full open and observe the function of damper to ensure that there is free, unobstructed operation through the entire angle of damper travel.

  - Pick a cool day, when the economizer damper is wide-open and warm the outdoor air temperature sensor with your hands or an electric hair dryer. The outdoor air damper should move to its minimum position. If not, either the sensor needs calibration or the economizer control is malfunctioning.

- Waterside economizer. When outdoor conditions are cool, waterside economizers use evaporative cooling (provided by a cooling tower) to produce cooled water. The cooled water is then used by waterside chillers to supply to an air conditioning system.

Other steps can be taken to reduce the cooling costs generated by different circumstances.

- Use the Sandia Cooler [6]. At the Energy Department's Sandia National Laboratories, researchers have developed an innovative new air-cooling technology—the Sandia Cooler—which improves the way heat is transferred in computers and microelectronics, significantly reducing the energy needed to cool CPUs in data centers. If the technology can be successfully scaled up and applied to other applications like heating, ventilation, and air conditioning, researchers say the Sandia Cooler has the potential to decrease overall electrical power consumption in the United States by more than 7%. To understand the technological advances in the Sandia Cooler, it helps to understand traditional CPU coolers. Comprised of a fan and a finned aluminum or copper heat sink, traditional CPU coolers have longstanding technological problems. Their effectiveness is hampered by the "boundary layer effect"—a thin layer of motionless air that adheres to the heat sink fins. This layer of "dead air" acts like an insulating blanket, trapping warm air. While the spinning fan blades of conventional CPU coolers remain relatively dust free, the stationary heat sink fins collect dust, which greatly reduces airflow and the CPU cooler's ability to convey heat to the surrounding air. With an innovative, compact design that solves the problems traditional CPU coolers face, it is no surprise that the Sandia Cooler is a recipient of R&D Magazine's 2012 R&D 100 Awards and one of three to receive R&D Magazine Editor's Choice Award. The Sandia Cooler combines a fan and a finned metal heat sink into a single element called a heat-sink impeller in which the fins of the heat sink rotate, improving heat transfer by a factor of ten. The unique geometry of the rotating fins provides for an extremely quiet operation, and the rapidly rotating fins minimize dust buildup. Supported by the Energy Department's

Building Technologies Program, the Sandia Cooler is a break-through technology with the potential to advance heat exchangers beyond what is available today. Heat exchangers are used to cool the surrounding air in microelectronics, refrigeration, heating, air conditioning, and nearly every application that generates waste heat—accounting for 53% of building energy consumption. While currently available for licensing for electronic cooling and LED lighting technology, researchers continue to refine the Sandia Cooler and are working on scaling up this new heat exchanger technology for applications such as heating and cooling systems.

- Air filters. Air filters play critical roles in maintaining indoor air quality and protecting downstream components of an air-handling system (e.g., the cooling coil and fan) from accumulating dirt that can reduce equipment efficiency. Dirty filters force air to go around filtration sections and unfiltered air deposits dirt on the cooling and heating coils rather than on the filter. It takes more effort and requires more skill to clean a dirty evaporator coil than to replace a filter.

    Solution: Routinely change filters based on the pressure drop across the filter, calendar scheduling, or visual inspection. Scheduled intervals should be between 1 and 6 months, depending on the dirt load from indoor and outdoor air. Measuring the pressure drop across the filter is the most reliable way of rating dirt loading on the filter. In facilities with regular and predictable dirt load, measuring pressure drop across the filter can be useful to establish appropriate filter-changing intervals. Thereafter, filter changes can be routinely scheduled. A complete air-filter pressure kit costs approximately $70. Hardware for installing the pressure taps costs less than $10. A service technician can do installation in just a few minutes.

- Evaporator and condenser coil. A dirty evaporator or condenser coil will reduce cooling capacity and compromise equipment energy efficiency. A clogged evaporator coil reduces airflow causing the compressor motor to consume more energy. When exposed to unfiltered outdoor air, condenser coils easily trap dust and debris causing condensing temperature to rise and reducing cooling capacity. A Pacific Gas & Electric (PG&E) study showed that a dirty condenser coil could increase compressor energy consumption by 30%.

Solution: Visually inspect the evaporator and condenser coils at least once a year, for clean airside passage. Replacing filters on a regular basis will keep the evaporator coil clean. Remove dirt from coils by washing and vacuuming. To ensure that coils are not damaged by high-pressure spray wash, an experienced cleaning crew should be engaged. In case of a two-person crew, one person sprays the coil clean, while the other person continuously vacuums out the cleaning solution.

- Refrigerant charge. Incorrect refrigerant charge can significantly affect a direct expansion (DX) unit's energy efficiency. A recent field study of 74 refrigeration systems showed that over 40 were incorrectly charged. Correcting the refrigerant charge can increase cooling cost reduction from 5 to 10%. Therefore, it is important to verify the correct refrigerant charge in your systems. Some signs indicating an undercharged system include frosting on the evaporator entrance, a warm suction line, a cool liquid line, warm air supply, and continuous compressor operation. In an overcharged system, excess liquid refrigerant backs up in the condenser and increases the head pressure.

    Solution: Determine if refrigerant charge is acceptable, insufficient, or excessive. Using the manufacturer's chart, look up the evaporating temperature that corresponds with the measured suction line pressure. Measure the actual suction line temperature. Determine the superheat as the difference between the measured suction line temperature and the evaporating temperature from the chart. For most direct expansion systems, a superheat temperature between 10°F and 20°F indicates adequate refrigerant charge. For systems with thermal expansion valves, the degree of sub-cooling should also be checked and compared to manufacturer's recommendations in order to determine if the system has the right refrigerant charge.

- Leaks in cabinet and supply ducts. Pressurized air can easily find its way through leaks in the unit cabinet and ductwork. Leaking air reduces cooling capacity and leads to loss of energy from cooled air. Based on a recent study of 350 small commercial HVAC systems in southern California, energy benefits from cabinet integrity and duct sealing are estimated at about 20% of the annual cooling consumption. Comfort in buildings with tight HVAC systems is expected to improve because the system will be able to deliver sufficiently cooled air (as designed) to serve the space loads.

> Solution: An easy task is to check the cabinets and correct air leakage. Some corrective actions include replacing screws or latches, patching or replacing gaskets, or replacing missing screws on loose access panels. Do not forget to recharge p-traps or u-bend water traps to condensate drain pans because the condensate drain pipe is another potential source of air leakage. Furthermore, duct leakage testing and sealing can be done, even though this will cost more resources and skills. Duct leakage testing should be conducted using the duct pressurization method described in the SMACNA2 Air Duct Leakage Test Manual. Aeroseal is a new technique that combines duck leakage testing and sealing into one operation.

- Condenser water temperature. Most chillers reach their maximum operating efficiency at the designed peak load. However, chillers operate at part-load condition most of the time. Resetting the condenser water temperature normally decreases the temperature lift between the evaporator and the condenser, thus increasing the chiller operating efficiency.

> Solution: Reset the condenser water temperature to the lowest possible temperature. Allow the cooling tower to generate cooler condenser water whenever possible. Remember that although lowering the condenser water temperature will reduce chiller energy, it may increase cooling tower fan energy consumption because the tower fan may have to run longer to achieve the lower condenser water temperature. In addition, some older chillers have condensing water temperature limitations. Consult the chiller manufacturer to establish appropriate guidelines for lowering the condenser water temperature.

- Import part load performance. Electrical chillers often represent the single largest electrical load in facilities, accounting for between 35 to 50% of a building's annual electricity consumption. If you have a multiple-chiller system serving your facility, make sure each chiller operates at no less than 50% of rated load (if at all possible). This eliminates the potential that the chiller will operate at partial load, causing low efficiency of equipment, as well as reduced equipment life resulting from unnecessary operation.

Solution: Consider staging multiple chillers, operate the fewest possible number of chillers required to meet the load, and use the highest overall operating efficiency. This can be achieved by using an electronic control system to calculate the actual cooling load and matching the right number of chillers needed for the particular load.

There are many more possibilities of cost-effective improvement of cooling operations, saving energy, and reducing electricity costs during the summer. These include:

- Raise thermostat settings for cooling, if applicable.

- Reduce cooling system run hours as much as possible.

- Reset chilled water temperature.

- Implement and improve water treatment in cooling towers.

- Clean evaporator and condenser tubes to remove scale or buildup.

- Clean fan blades, lubricate bearings, and adjust belts.

- Minimize the use of reheat as best as possible.

- Commission ventilating systems through testing, adjusting, and balancing.

- Ensure that control valves operate correctly.

- Optimize multiple pump controls to reduce pumping operating costs.

- Do not cool unused space.

1.5 END OF LIFE OF COMPUTER SYSTEMS/COMPONENTS

The last phase in the lifecycle of computer systems or components is referred to as the end-of-life phase. At this stage, they are out of use or no longer meet the needs of their owners/operators, who usually want to get rid of them. The focus at this point is how to properly dispose of these components as well as what subsequently happens to them.

1.5.1 Alternatives to Recycling

In many cases, the recycling of old computers can be avoided. Some practical examples are:

- Just because a computer is obsolete for the needs of one individual does not mean that it cannot be useful to someone else. There are several donation programs that will take over the computer ownership, get it in working order, and give it to someone who needs a "starter machine" to get into the information technology age. Sometimes it is not necessary to replace a whole computer. It may be sufficient to exchange selected components (hard drive, processor, memory) and improve the performance of an "old" computer.

- If the "old" computer is too old to support a recent operating system, it is often possible to install a light version of an operating system, so that the computer can be used for activities like web browsing, simple word processing, or other low performance tasks.

- In some cases, selling old components may also be an option.

1.5.2 Disposing Old Computers

If no other alternative to recycling is applicable for the end user, then the old device has to be disposed. In this state, it is electronic waste. With an expanding global market for electronic goods and the lifespan of many products getting shorter, there is rapid growth in electronic waste (e-waste). The UN estimates that 20 to 50 million tonnes of e-waste are globally produced each year [8].

There are many alternative ways of getting rid of old electronic products. For example, they can be returned to the seller, they can be picked up from the office by a recycler, or they can be dropped off at specialized collection sites. However, getting rid of old components does not necessarily mean that they will be processed in a responsible manner (a responsible recycler makes sure that all aspects of the business are managed as [ecologically] safely as possible).

E-waste or e-junk can be disposed of in several ways:

- Computer components disposed in landfills are problematic due to potential contamination of soil and water from leaking toxins. Although there is current disagreement regarding the plausibility of this concern, many European countries have already banned the practice. The problem remains acute in developing countries, however, where people are more likely to live in close proximity to a landfill.

- Burning old computers releases heavy metals and highly toxic fumes into the air. This can cause respiratory and skin problems to exposed persons, adversely affecting their health. This typically happens in developing countries, where there is no possibility for collecting and recycling e-waste.

- Since electronics may still contain valuable materials, they could be recycled. In developing countries, where cost of recycling is often low, this is usually done by hand, exposing workers and neighborhoods to adverse health risks.

Very often, recycled materials are sold to brokers who ship them to developing countries, most frequently to China [9], which has become a dumping ground for international e-waste or e-junk. Domestic generation of e-waste is also a growing challenge in China and the Chinese government has started to take measures as dangers (ecological and human health-related consequences) associated with improper disposal of electronic equipment are becoming more and more apparent. There is also considerable international discussion on stricter regulation of e-waste trade. Similar to the practice of ship breaking, the import of e-waste is generally aided by intermediaries in Hong Kong, the Philippines, and other Asian countries. According to a Worldwatch Institute report, discarded equipment is shipped to Hong Kong in containers labeled "for recycling," and then smuggled to several "recycling towns" in the adjacent Guangdong Province of China. From there, much electronic waste is outsourced to farmlands for dismantling. In September 2006, authorities from Guangdong Province and Guangzhou's Haizhu District uncovered dozens of illegal e-waste processing workshops in one of the biggest lawsuits in the province. Forty tons of e-waste was headed for the huge secondhand IT markets in the Pearl River Delta and, according to local authorities, contained over 350 harmful chemical materials. The Chinese news media is increasingly covering crackdowns on such illegal e-waste operations.

During the last years, trade with e-waste has spread to other regions in the world, particularly to West Africa. Sending old electronic equipment to developing countries is often hailed as "bridging the digital gap." However, all too often this simply means dumping useless equipment on the poor. One estimate suggests that 25 to 75% of "secondhand goods" imported to Africa cannot be reused [8].

To regulate international e-waste trade, the Basel Convention on the control of trans-boundary movements of hazardous wastes and their disposal (which is the most comprehensive global environmental agreement on hazardous and other wastes) came into effect in 1992. The convention was signed by 175 parties and aims to protect human health and the environment against adverse effects of the generation, management, and trans-boundary movement and disposal of hazardous and other wastes.

1.6 HOW CAN IT POSITIVELY INFLUENCE THE ENVIRONMENT?

Thanks to IT infrastructure and IT software, it is now possible to achieve a lot in an ecologically friendly way. Let us have a look at some examples:

- Less use of paper. You would be surprised at how much paper lands on your office desk, which is destroyed after reading. Electronic documents are preferable and have a positive impact on the environment. Instead of having specific documents delivered to their desks, workers could get links to these documents and decide if they need to print them out. As such, waste of paper is prevented. It is possible to restrict access to documents using common firewalls and intranet technologies, and only those who are connected to the organization's network can view them.

- Remote instead of on-site support. For maintenance purposes or in case of related problems (i.e., for software or hardware organizations), in the past it was necessary to do on-site support in order to solve customer-related problems. Thanks to the Internet and suitable software (e.g., Teamviewer, VNC) it is possible today to connect to a remotely located computer to gain complete access to or control over the device. This could help in better error analysis and reduce or eliminate transportation costs. In addition, the transportation time is no longer a problem as direct support is possible (as long as the network is available).

- Online meetings. Let us assume that several people in different countries are working on a common project and want to discuss the project status. Having all participants travel to a single location would mean travel costs, travel duration, and emission into the environment. Thanks to Voice over Internet Protocol (VoIP) and related network infrastructure, it is possible to have a telephone conversation or conference over the Internet. This would suffice for simple discussion needs. Suitable software (Skype, etc.) also offer a video function and can be used if need be. Sometimes there is need to do more than just speaking (e.g., to review documents for presentations, etc.). In this case, other tools like Mikogo, which allow each participant to share his or her desktop (which is displayed to the other participants), can be relied upon.

- Use energy-efficient hardware. Energy efficiency is an important issue and it should be kept in mind when acquiring hardware. It is advisable to look for machines that are made with energy-efficient components.

- Design products to last instead of designing products for the dumpsite. Many electronic products are designed for the dumpsite [10]. They have short life spans, are often expensive to repair, it is sometimes difficult to find spare parts, and they easily become obsolete. Many consumer-grade electronics products are cheaper to replace than to fix, even if someone who can fix them can be found. Because they are designed using many hazardous compounds, recycling these products involves processing toxic material streams, which is never 100% safe. Before any electronics company makes the claim of being green and sustainable, it must first go from producing "disposable" products with limited life spans (planned obsolescence) to products that are designed to last (sustainable). Instead of purchasing products with high failure rates that need frequent replacement, consumers should choose long-lasting, upgradeable goods, with long warranties, that can be efficiently repaired and recycled.

1.7 CONCLUSION

Previously, we described how each stage of a computer system's lifecycle impacts the environment. We also presented some measures that could help reduce or prevent the negative impact of IT. However, the whole process is not trivial and requires collective effort from several parties:

- Manufacturing and purchasing

  - Manufacturers. Change in manufacturers' philosophy is needed. "Designing for the dumpsite" may be attractive for manufacturers because they are interested in selling their products and want the customers to keep buying. At least using nontoxic materials during production would prevent health problems during recycling.

  - Customers. If all customers decided to buy only products having a valid, legal, and controlled eco-label, manufacturers would be forced to change their philosophy.

  - Legislation/state/organizations. In the chaos of many existing eco-labels, customers need to be well informed in order to know valid labels and avoid "green washing."

- Use. In corporate settings, especially, we have seen that major challenges in computer infrastructure use are power consumption and cooling. Very often purchasing appropriated hard- or software prevents problems related to power consumption and high costs in the immediate future. Here it is important always to keep the bigger picture in mind.

- Disposal. Developing countries are mostly target dumpsites for old computers. Reasons for this include the following:

  - There are no laws against the import/export.

  - The export is illegal or masked as reuse/donation.

  - The export is illegal or under cover of recycling.

  - The export occurs under trade agreements.

- Here there is a need for the local institutions and individuals in developing countries to change their mentality and not allow or engage in the illegal import of toxic trash from other countries and foster the implementation of legislations that will forbid dumping across borders.

REFERENCES

[1] Velte, T., Velte, A., and Elsenpeter, R. 2008. *Green IT: Reduce Your Information System's Environmental Impact while Adding to the Bottom Line.* Velte Publishing, Inc.

[2] Department of Health, State of New York. 2009. Lead Exposure in Adults—A Guide for Health Care Providers. http://www.health.state.ny.us/publications/2584/ (accessed February 4, 2013).

[3] Bridgen, K., Webster, J., Labunska, I., and Santillo, D. 2007. Toxic Chemicals in Computers Reloaded. Greenpeace's Research Laboratories Technical Note 06/07.

[4] Kailas, S.V. 2007. Economic, environmental and social issues of material usage. Lecture "Material Science," Chapter 18.

[5] Meier, A.K. and LeBot, B. 1999. One Watt Initiative: A Global Effort to Reduce Leaking Electricity. Lawrence Berkeley National Laboratory.

[6] Matulka, R. 2012. Sandia Cooler Blows Traditional CPU Coolers Away. http://energy.gov/articles/sandia-cooler-blows-traditional-cpu-coolers-away (accessed February 4, 2013).

[7] Federal Energy Management Program. 2005. Actions You Can Take to Reduce Cooling Costs. Fact Sheet.

[8] Greenpeace. 2008. Poisoning the poor – Electronic waste in Ghana.

[9] Jones, S.L. 2007. China as E-Waste Dumping Ground: A Growing Challenge to Ecological and Human Health.

[10] Electronics TakeBack Coalition. 2010. Designed for the Dump. http://www.electronicstakeback.com/designed-for-the-dump/ (accessed February 4, 2013).

CHAPTER 2

Green Computer Science

Methodologies, Designs, Frameworks, and Tools That Can Be Used to Compute Energy Efficiently

DEFINITIONS AND ABBREVIATIONS

C: Programming language initially developed by Dennis Ritchie

C++: Intermediate level programming language

CPU: Central processing unit

GHz: 1.000.000.000 Hz

Hz: SI unit for frequency

I/O: Input/output

HIFI: High fidelity

RAM: Random access memory

ROM: Read only memory

SI: International System of Units (abbreviated SI from French: Le Système national d'unités)

2.1 INTRODUCTION

Nowadays, almost every system including mobile devices, HIFI equipment, data centers, wireless communication systems, and software systems that run on computers or other systems need energy to work and are managed using software. That is why energy today is a very important factor in the design and development of software. Computer science that better understands the properties of programs used to manage software will go a long way in reducing energy consumption. Energy efficient computing is currently used today in several domains. These include:

- The development of large data centers for supporting web-based services such as search engines, email, online shopping, etc.

- The development of mobile and embedded devices (the extension of battery time is a big challenge in this case).

- The development of wireless communication.

- The development of computer games.

In this chapter, we analyze methodologies, concepts, and tools in computer science that can be used to improve energy efficiency in software.

2.2 ABOUT COMPUTER SCIENCE

Computer science is the study of computers and computational systems. This includes theory, design, development, and application. Principal areas in computer science are artificial intelligence, computer systems, database systems, human factors, numerical analysis, programming languages, software engineering, and the theory of computing. Computer science incorporates concepts from mathematics, engineering, and psychology.

A computer scientist is occupied with solving issues that range from abstract challenges to practical solutions. The main challenge here is to determine what problems can be solved using computers and the complexity of algorithms needed, and the practical solution here is the design of computer systems for easy human use. Computer scientists build different computational models of systems ranging from physical phenomena (condensation cloud, weather forecasting, etc.) to human behaviors (reasoning, mechatronic, robotics, etc.) and computer systems themselves (system monitoring, performance evaluation, etc.). Such models often require extensive numeric or symbolic computation. Computer scientists design and analyze algorithms used in solving problems as well as develop and study the performance of computer hardware and software.

The complexity of algorithms is one of the main factors that need to be considered in determining energy efficiency of software.

In mathematics, computing, and related subjects, an algorithm is an effective method based on a finite set of unambiguous instructions

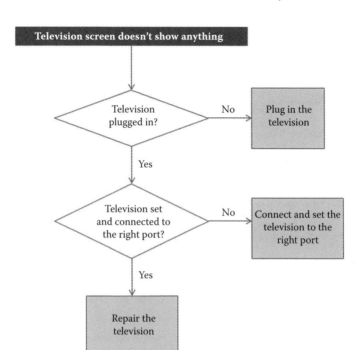

FIGURE 2.1 This algorithm aims at determining why the television does not show anything and proposes a stepwise solution for this problem.

performed in a prescribed sequence for solving a problem. A graphical representation of an algorithm is shown in Figure 2.1. This algorithm aims at determining why the television screen does not show anything and proposes a stepwise solution for this problem (primitive instructions).

2.3 ENERGY EFFICIENCY IN COMPUTER SCIENCE

Energy efficiency (the efficient use of energy) means reducing the amount of energy required to provide products and services while maintaining or improving their quality. In computer science where software is used to solve problems, the goal is to use efficient algorithms that help reduce the energy consumption while solving problems to an acceptable level or better still to full precision.

Let us recapitulate some basic knowledge about computer hardware. Figure 2.2 presents an example of computer system architecture. In this figure, the components are interrelated.

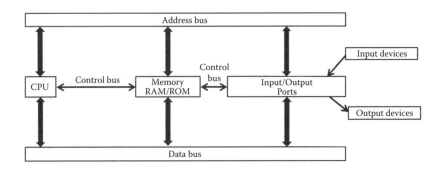

Figure 2.2   Example of computer system architecture.

2.3.1 The Central Processing Unit (CPU)/Processor

Processors form the heart and soul of any computing system. They carry out the instructions in a given computer program, which typically involves performing some operation on a given data set. Operation types can be arithmetic, logical, or input/output.

All processors contain some basic building blocks. The implementation and the design of these blocks make the difference between the processors. Every processor supports a set of instructions, which can be used in a program to perform different operations.

A computer program is basically written code in a high-level programming language (which is understandable to humans) like e.g. Java or C/C++. In order to be understood by the processor, this written (source) code has to be converted into a machine code. This is the compiling process. Figure 2.3 shows a simplified description of the compiling process of a program written in C: The source code (e.g., "program.c") is first preprocessed. This means that some functions are applied to the source code before the compiler

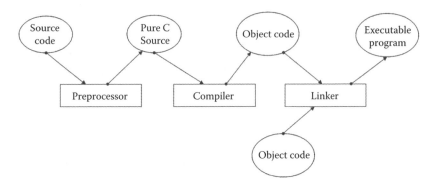

FIGURE 2.3   The simplified compiling process of a C program.

processes it. After this step, the final source code is checked for syntax and semantic errors, for example. If no errors occurred during the compiling process, object code (e.g., "program.o") is generated. The object code can be linked with other object codes (e.g., from archives) and the final output is the executable program file (e.g., "program").

Once compiled, a program can be loaded onto a memory device and executed by a processor. Nowadays, software can become very huge and the end user mostly expects the machine to be fast and produce accurate results using less time. This may not be an appropriate example (due to the current economic situation in the banking system), but if day traders had to wait too long for their risk management system to give them risk measures, they would not be able to react quickly on the market and make important buy/sell decisions promptly. In order to be faster, which means process more instructions at the same time, also called parallel processing, it is obvious that one CPU will not be enough. You will need more processing units. There are several alternatives for this purpose:

- Multiprocessor systems: computer systems having several CPUs that are using their own sockets, are independent from each other, and have direct access to the motherboard through their own set of pins.

- Multicore systems: computer systems having one CPU. The CPU itself contains several processing units but they all fit in one socket, which allows them to share the Level 1 cache, for example.

2.3.2 Memory (RAM and ROM)

In the computing world, memory refers to the physical devices used to store programs or data on a temporary or permanent basis for use in a computer or other digital electronic device. Before an instruction is executed, it is stored at a particular address in the memory. Once the execution is over, the data can be written back to another address in the memory.

Every computer comes with a certain amount of physical memory, usually referred to as main memory or RAM (random access memory). You can both write data into RAM and read data from RAM. This is in contrast to ROM (read only memory), which permits you only to read data. Most RAM is *volatile*, which means that it requires a steady flow of electricity to maintain its contents. As soon as the power is turned off, whatever data was in RAM is lost. Computers usually contain a small amount of ROM that holds instructions for starting up the computer.

2.3.3 Input/Output (I/O) Ports

External I/O ports are provided on the main board to enable the system to be connected to peripheral devices such as some input devices (mouse, keyboard, scanner) and output devices (printer, video display unit, audio speakers).

2.3.4 Bus

The bus is a subsystem or a route that is used for the transfer of data between modules in computer systems. All buses consist of two parts—an address bus and a data bus. The data bus transfers actual data, whereas the address bus transfers information about where the data should go.

The size of a bus, known as its *width,* is important because it determines how much data can be transmitted at one time. For example, a 16-bit bus can transmit 16 bits of data, whereas a 64-bit bus can transmit 64 bits of data. Data bus width determines the overall system performance and typically consists of 8, 16, 32, or 64 parallel lines. Address bus is used to indicate the location of the source or destination of data transfer process.

These components need power in order to run and provide the expected results. Depending on how a program is written, the effects on the power consumption can be different. Let us illustrate this difference with a practical example.

We often need to sort data because we want to display them in a specific order or we need to access them quickly. Several methods exist to solve the sorting problem. The easier approach would be a bubble sort algorithm and a more complex approach is the quick sort algorithm.

The bubble sort algorithm consists of the following steps [1]:

- Comparing each adjacent pair of items in a list in turn

- Swapping the items if necessary

- Repeating the pass through the list until no swaps are done

Complexity is $O(n^2)$ for arbitrary data, but approaches $O(n)$ if the list is nearly in order at the beginning. Bidirectional bubble sort usually does better because at least one item is moved forward or backward to its place in the list with each pass.

Bubble sort can be used to sort a small number of items (where its asymptotic inefficiency is not a high penalty). It can also be used on a list of any length that is nearly sorted.

The quick sort algorithm consists of the following steps [1]:

- Pick an element from the array (the pivot).

- Partition the remaining elements into those greater than and less than this pivot.

- Recursively sort the partitions.

There are many variants of this basic scheme: to select the pivot, to partition the array, to stop the recursion on small partitions, etc.

Quicksort has running time $O(n^2)$ in the worst case, but it is typically $O(n \log n)$. In practical situations, a finely tuned implementation of Quicksort beats most sort algorithms, including sort algorithms whose theoretical complexity is $O(n \log n)$ in the worst case.

In other words, it is necessary to use the right (and efficient) algorithm in order to write efficient software.

2.4 PROFILING—ASSESSING SOFTWARE EFFICIENCY

Several issues in software development occur because of inefficient programming. Because software program efficiency also affects its power consumption, the possibility of finding out which part of your program takes too long is a big help in the way of writing efficient software.

In order to make such helpful findings, you will have to analyze your software. You can analyze the program's source code or you can observe the software when it is running. The process of analyzing a program while it is running is called profiling and is the subject of this section.

While the program is running, several measures, which can be examined after the program's execution, are made. With the help of such measurement, it is possible to answer different questions such as:

- How long was the program executed?

- How often was a special routine called?

- What was the average execution time of a specific function?

- In which order were several functions called?

Profiling is an important step when assessing software efficiency. If you can answer the previous questions (and more), you can decide later what you can make differently in order to make your program more efficient regarding specific goals.

2.4.1 Profiling Methods

The method used to gather runtime information about computer programs could be grouped into two categories:

- Noninvasive methods

- Invasive methods

2.4.1.1 Noninvasive Profiling

A noninvasive profiling method is a profiling method where the observed software is not modified before its execution. An example of noninvasive profiling is the sampling. The sampling process usually works as follows.

The observed software (subject) is launched together with an observer software. The observer monitors the subject and is responsible for performance measurements, for example. The observer calls periodically an interrupt (short interruption of a running program in order to run an interrupt routine) and reads the content of the program counter. Then the observer looks up the stack in order to find out which procedure is being called. Finally, the observer increments the snapshots counter of the called procedure.

If a function f is found consecutively in n snapshots, based on the snapshot frequency it is possible to make a statistical assertion about the execution time of the function f. If a function is called more than once, you can also make statements about its average execution time using the same procedure.

Pros:

- The sampling process does not make any changes in the observed program (subject).

- The subject's source code does not need to be modified.

- It is possible to make software and system profiling (observe system routines) as well.

- The observer's influence in the subject's execution time is minimal.

- By selecting the suitable sampling frequency, it is possible for example to only observe routines having a specified execution time. This can be useful to find out if execution time is wasted by a single function.

Cons:

- Sometimes the subject has to be rebuilt in order to enable profiling.

- Observation's results are not exact because they rely on a statistical evaluation.

- It is possible to generate a call graph using the profiling results. However, the call graph also relies on a statistical evaluation.

2.4.1.2 Invasive Profiling

In opposition to the noninvasive profiling method, the observed subject needs to be modified before the analysis. The modification consists of adding several calls to observation routines within the program. This process is called instrumentation. There are several types of instrumentation, which differ in the moment when the instrumentation is done.

However, because of the instrumentation the observed program becomes slower. Then the execution of profiling functions also costs CPU time. The more profiling routines are called, the bigger is the overall execution time overhead.

2.4.1.2.1 Source Code Instrumentation   This type of instrumentation occurs before the subject is compiled (Figure 2.4) and is usually processed by the developer. The developer has to retrieve performance information. In order to do this, the developer makes different calls to special function at several places in the subject's source code.

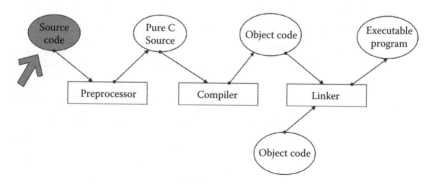

FIGURE 2.4   Instrumenting the source code.

```
long timeBefore = System.currentTimeMillis();
doSomeJob();
long timeAfter = System.currentTimeMillis();
long duration = timeAfter - timeBefore;
```

FIGURE 2.5 Example of simple source code instrumentation.

Figure 2.5 shows how the execution time of a specific function can be retrieved. The developer just memorizes the system time before and after the execution of the specific function.

This type of instrumentation comes with the following problems:

- In operating systems with preemptive scheduling strategies, it is possible for a process to be interrupted by the scheduler because another process having a higher priority must be run. In this case, the calculated overall execution time will also include the execution time of the higher prioritized process (or at least a part of it).

- If a thread (light-weighted process) is created within the observed function (doSomeJob()), the execution time of the main process will not include the execution time of the child thread. If the created thread is responsible for a bottleneck, you will not be able to find this out using this instrumentation type. Alternatively, you could also add observation routines at the beginning of the child's entry point (and at the end as well).

This instrumentation type is adequate for small programs. One condition for using this method is that you already have an idea about which function is time intensive and must be observed. Otherwise, you will have to manually insert observation routines into every function, which makes the search for inefficient code parts harder.

Another condition for this instrumentation type is that you have to modify the source code in order to perform the analysis. This is not always wanted.

2.4.1.2.2 Instrumentation at Compile Time   This type of instrumentation is processed by the source code compiler (see Figure 2.6). Some code compilers allow the activation of profiling routines when special compiler options are enabled. For example, if you use the GNU C compiler (gcc) or the GNU C++ compiler (g++) with the option "–pg", the compiler will add profiling routines to the executable. After the subject's execution, a file

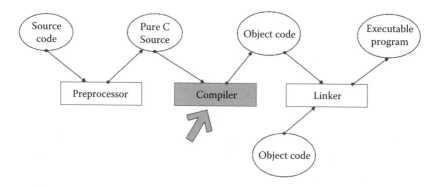

FIGURE 2.6   Instrumenting at compile time.

gmon.out is generated. The file can be processed via diverse tools in order to gather specific information.

2.4.1.2.3 Object Code Instrumentation   This type of instrumentation happens after the subjects have been compiled and before the link process (see Figure 2.7). With this method, the object files are patched and altered with call to analysis routines, for example, before each function call and after each function return. In this way, the original code is extended with extra code allowing the gathering of reliable analysis data.
    Pros:

- The object code can be automatically instrumented using specific tools.

- This method delivers reliable results (compared to sampling).

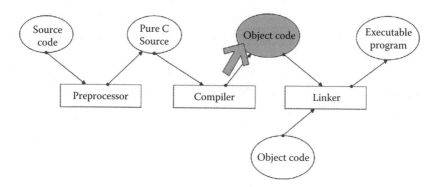

FIGURE 2.7   Instrumenting object files.

Cons:

- The source code has to be rebuilt.

- The instrumented program runs slower.

2.4.1.2.4 Instrumentation at Runtime   A further profiling technique is the simulation. Here the subject is not executed but simulated in a model of the target platform. In other words, the source code instructions are not directly executed by the CPU. Instead, they are processed by the target platform's simulator. A detailed target platform's model leads to accurate observations.

Because the subject is simulated, there is no need to rebuild the subject's source code.

2.4.1.2.5 Sampling vs. Implementation   The results of a sampling-oriented profiling method are not exact because they are based on statistical analysis. It is possible to repeat the sampling procedure more than once for a better approximation of the measurement results. However, if exact results are needed, it is better to use an instrumentation-based profiling method.

It is not practical to instrument a whole source code because of the observer's influence in the instrumented subject. When possible, it is better to first run a sampling-based profiling in order to detect time intensive routines at an early state. Then you can instrument the specific functions (and only those) in order to gather accurate runtime information.

2.4.2 Code Profiler

Now that we addressed different methods that allow us to gather runtime information about software, let us get more concrete and describe tools implementing those profiling methods. Such tools are called code profilers.

Profilers are mostly developed to support specific programming languages and operating systems. We will restrict the scope of the description to Linux-based profilers, specifically gprof, OProfile, and Valgrind.

2.4.2.1 Case Study: gprof

The GNU Profiler (gprof) exists in most Unix distributions because it is included in the GNU binutils package. Gprof is used with the GNU C compiler (gcc) or the GNU C++ compiler (g++). The profiling process consists of three steps (see Figure 2.8):

```
[lionel@benq]$
[lionel@benq]$ls
CallGraphNoThread.cpp
[lionel@benq]$g++ -g -pg CallGraphNoThread.cpp -o callgraph-nothreads
[lionel@benq]$ls
CallGraphNoThread.cpp  callgraph-nothreads
[lionel@benq]$./callgraph-nothreads
[lionel@benq]$ls
CallGraphNoThread.cpp  gmon.out  callgraph-nothreads
```

FIGURE 2.8   Profiling using gprof.

1. First, the subject is compiled with gcc or g++ with the flag "-pg" and linked. With this option, additional code (for the output of profiling data) is generated and added to the subject. You can also use the option "-g" in order to have debug information (e.g., symbol names) in the output profiling data.

2. Then the subject is executed as usual and at the same time profiling data are collected. At a successful termination (with exit or return), the file "gmon.out" is generated.

3. Finally, the profiling data are fetched from the file "gmon.out" and analyzed using gprof.

gprof supports three profile types:

1. Flat profile: a tabular listing of functions with corresponding profiling data. The flat profile in Figure 2.9 can be interpreted as follows: The function named "callgraph" has an average execution time of 6.03 seconds and was called twice during the program execution. The execution time of this function is the highest and corresponds to 29.18% of the main program's execution time.

```
[lionel@benq]$gprof callgraph-nothreads -p -b
Flat profile:

Each sample counts as 0.01 seconds.
  %   cumulative   self              self     total
 time   seconds   seconds    calls  s/call   s/call  name
29.18     3.50      3.50        2    1.75     6.03   callgraph()
28.60     6.93      3.43        2    1.71     4.28   methodA(bool)
28.51    10.35      3.42        2    1.71     2.56   methodB(bool)
14.26    12.06      1.71        1    1.71     1.71   methodC()
[lionel@benq]$
```

FIGURE 2.9   Flat profile with gprof.

2. Call graph: a tabular display of functions showing in which order functions where called. Figure 2.10 shows an example of call graph profiling with gprof. At the index [4] you can see method() was called by method() and calls itself methodC(). For a better visualization of the callgraph, the tool kprof is more adequate (see Figure 2.11).

3. Source code annotation: This is a tabular display of the subject's source code with relevant profiling data. Figure 2.12 shows an example of annotated source using gprof. Near each function in the source code, you can see how often the function was called.

gprof works best with single-threaded programs. The following scenario will occur if you compile a multi-threaded application with "-pg" and run it: at the program termination, each thread/process will overwrite the file "gmon.out" and at the end, only the file generated by the last process/thread will be available. A known workaround for this issue consists of setting the environment variable GMON_OUT_PREFIX. In this way, each process will generate a file "gmon.out.pid" (where pid is the id of

```
[lionel@benq]$gprof callgraph—nothreads —c —b —no—flat—profile
                             Call graph

granularity: each sample hit covers 2 byte(s) for 0.08% of 12.06 secs
index % time    self  children    called     name
                                                   <spontaneous>
[1]     100.0   0.00   12.06                    main [1]
                3.50    8.56        2/2             callgraph(void*) [2]

                3.50    8.56        2/2         main [1]
[2]     100.0   3.50    8.56        2         callgraph(void*) [2]
                3.43    5.13        2/2             methodA(bool) [3]

                3.43    5.13        2/2         callgraph(void*) [2]
[3]      71.0   3.43    5.13        2         methodA(bool) [3]
                3.42    1.71        2/2             methodB(bool) [4]

                3.42    1.71        2/2         methodA(bool) [3]
[4]      42.5   3.42    1.71        2         methodB(bool) [4]
                1.71    0.00        1/1             methodC() [5]

                1.71    0.00        1/1         methodB(bool) [4]
[5]      14.2   1.71    0.00        1         methodC() [5]

Index by function name
   [3] methodA(bool)              [5] methodC()
   [4] methodB(bool)              [2] callgraph(void*)
[lionel@benq]$
```

FIGURE 2.10 Call graph display with gprof.

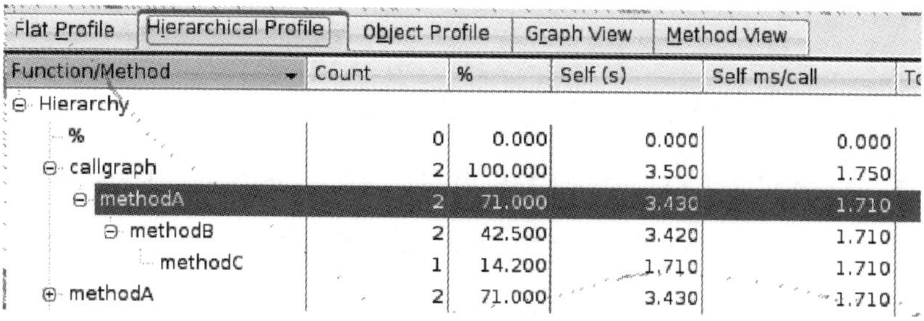

Function/Method ▾	Count	%	Self (s)	Self ms/call	T
⊖ Hierarchy					
%	0	0.000	0.000	0.000	
⊖ callgraph	2	100.000	3.500	1.750	
⊖ methodA	2	71.000	3.430	1.710	
⊖ methodB	2	42.500	3.420	1.710	
methodC	1	14.200	1.710	1.710	
⊕ methodA	2	71.000	3.430	1.710	

Tabs: Flat Profile | Hierarchical Profile | Object Profile | Graph View | Method View

FIGURE 2.11   Using kprof to visualize the gprof output.

the corresponding process), which will allow the separate analysis of each output file. This workaround applies only to processes (created using fork) and not to threads (created using pthread_create).

Pros:

- Free.

- Supports call graph profiling.

- Sampling-based profiler → minimal observer influence.

- Supports source code annotation.

- Graphical report display possible with kprof.

```
[lionel@benq]$gprof callgraph —nothreads —annotated—source
*** File /home/lionel/programming/cpptest/eval/CallGraphNoThread.cpp:
                #define MAX 500000000
                void methodA(bool callMethodC);
...

    2 -> void* callgraph(void* params)
                {
                        for(long i=0;i<MAX;i++);
                        bool *callC = (bool*)params;
                        methodA(*callC);
                        return (void*) 0;
                }
...

    1 -> void methodC(){
                        for(long i=0;i<MAX;i++);
                }
    1 -> int main(int argc, char* argv[])
                {
                        ...
                }
...
[lionel@benq]$
```

FIGURE 2.12   Annotated source display with gprof.

Cons:

- Cannot profile multi-threaded applications.
- The sampling frequency (100 samples per second) cannot be changed.
- Sampling → calculated execution times are not exact.
- Does not support kernel profiling.
- Source code annotation works for functions but you cannot get profiling data at a code line level.

2.4.2.2 Case Study: OProfile

OProfile is a collection of several tools that can be used for code profiling. For gathering profiling data, it is possible to choose between two methods:

1. Time-based sampling. A software clock calls an interrupt periodically and triggers the sampling.

2. Event-based sampling. When the profiler is started, the CPU counter is initialized with a predefined value and counts backward. An interrupt is called when the counter reaches 0 and triggers the sampling routine. At the end of the sampling procedure, the CPU counter is initialized again and the process is repeated.

OProfile consists of a kernel driver [2] and a daemon, which is responsible for the collection of profiling data. While the daemon is executed, information about all running processes in the user space and the kernel space as well are collected. For this reason, you need root permissions in order to run the profiler. Figure 2.13 shows an example of how OProfile is configured and started. First, the program to be observed is compiled (line 3). Then you can clear all existing profiling data (line 6). Before running the profiler for the first time, you need to select the sampling type. For event-based sampling, you need to run the command "opcontrol" and specify the path to the kernel image (lines 7 and 8). This allows the profiler to recognize which events are supported by the CPU. It is also possible to specify an event using the option "-event = EVENT_TYPE." Otherwise, the default event is selected. For time-based sampling, you just have to set the flag "-novmlinux." Then you can start the profiler.

```
 1  [lionel@benq]$ls
 2  CallGraph2Threads.cpp  CallGraphNoThread.cpp  reports
 3  [lionel@benq]$g++ CallGraph2Threads.cpp -o callgraph-multi -lpthread
 4  [lionel@benq]$ls
 5  CallGraph2Threads.cpp callgraph-multi CallGraphNoThread.cpp reports
 6  [lionel@benq]$sudo opcontrol --reset
 7  [lionel@benq]$sudo opcontrol
 8   --vmlinux=/home/.../kernel_src/linux-source-2.6.22/vmlinux
 9  [lionel@benq]$sudo opcontrol --start
10  Using default event: CPU_CLK_UNHALTED:100000:0:1:1
11  Using 2.6+ OProfile kernel interface.
12  Reading module info.
13  Using log file /var/lib/oprofile/samples/oprofiled.log
14  Daemon started.
15  Profiler running.
16  [lionel@benq]
```

FIGURE 2.13   Configure OProfile.

After the observed program terminates or whenever you think you have collected enough data, you can stop profiling (Figure 2.14, line 3).

To display a report of the collected data, you need to run the tool "opreport." The generated report contains the following data:

- Samples: Number of snapshots taken during the profiling session

- %: percentage count

- App name: the application name

- Symbol name: name of the analyzed function (available if the corresponding program was compiled with debug information)

As you can see in Figure 2.15, the actual execution time is not available because OProfile is sampling-based. However, this value can be approximated (only in case of event-based sampling) like this:

CPU Speed: 1.733 GHz = 1730 000 000 Hz × 1730000000 CPU ticks per seconds.

```
 1  [lionel@benq]
 2  [lionel@benq]$./callgraph-multi
 3  [lionel@benq]$sudo opcontrol --shutdown
 4  Stopping profiling.
 5  Killing daemon.
 6  [lionel@benq]$
```

FIGURE 2.14   Terminating the profiling session.

```
[lionel@benq]$opreport -1 | head -n15
CPU: PIII, speed 1733 MHz (estimated)
Counted CPU_CLK_UNHALTED events ... count 100000
samples  %        app name             symbol name
70241    27.9787  callgraph-multi      methodB(bool)
70238    27.9775  callgraph-multi      callgraph(void*)
70217    27.9691  callgraph-multi      methodA(bool)
35120    13.9891  callgraph-multi      methodC()
718      0.2860   vmlinux              dma_memcpy_pg_to_iovec
643      0.2561   libc-2.6.1.so        (no symbols)
445      0.1773   oprofiled            (no symbols)
239      0.0952   reiserfs             (no symbols)
230      0.0916   libqt-mt.so.3.3.7    (no symbols)
190      0.0757   oprofile             (no symbols)
179      0.0713   vmlinux              inflate_fast
178      0.0709   bash                 (no symbols)
[lionel@benq]$
```

FIGURE 2.15   OProfile report.

A sampling event is triggered each 100,000 ticks.
One sample is merely equivalent to 0.057 microseconds.
Pros:

- Free.

- Supports call graph profiling [3].

- Supports a large range of processors.

- Can be used in embedded Linux devices.

- Supports kernel profiling.

- Can handle multi-threaded profiling.

- 1 to 3% profiling overhead [3].

- Event-based and time-based sampling is possible.

- Source code annotation feature.

- Can compare the report of different profiling sessions.

Cons:

- Root permissions are needed.

- High configuration overhead if you have to rebuild the kernel.

- Supports a large range of processors.

- Execution time must be approximated.

2.4.2.3 Case Study: Valgrind

Valgrind is a package of tools that can (beside other things) do profiling and help you optimize your program. Valgrind emulates the CPU. In this way, Valgrind is able to check each memory access. However, an overhead in the execution time is observed because instructions are simulated. This behavior is illustrated in Figure 2.16. A simple multi-threaded test program is executed in 0.34 seconds (line 3). If this same program is simulated (without instrumentation) with Valgrind, the overall execution time is 9957 seconds (about 26 times slower!).

For call graph and execution time profiling, you need the tool callgrind. At the end of the simulation, a file callgrind.out.processid is generated, which can be visualized with the tool KCachegrind.

Pros:

- Free/open source.

- No need to rebuild the subject's source code.

- Multi-threaded profiling.

- Graphical report display possible with KCachegrind.

- Can do more than profiling.

```
1  [lionel@benq]$time ./callgrah-multi
2  done
3  real     0m0.340s
4  user     0m0.328s
5  sys      0m0.004s
6  [lionel@benq]$time valgrind —tool=callgrind ./callgrah-multi
7  .. valgrind output ...
8  done
9  .. valgrind output ...
10 real     0m9.057s
11 user     0m8.981s
12 sys      0m0.032s
13 [lionel@benq]$
```

FIGURE 2.16 Simulation overhead with Valgrind.

Cons:

- Huge simulation overhead (factor 26 in our example).

2.4.3 Optimization Best Practices

Since every developer can face optimization issues anytime, let us summarize the steps to follow in order to address a given optimization problem.

- Search and localize the bottlenecks. If the bottleneck is known, you can proceed to the next step. Otherwise, it would be helpful to use a code profiler to locate potential bottlenecks. You can also compare the results of different profilers to get a more accurate overview.

- Analyze, optimize, and test the inefficient areas. If needed, you can implement special unit tests responsible for testing the modified areas only. In this way, you can gain a better feeling about the impact of your changes.

- Apply the modifications to all the affected places and check the optimization results. Optimization issues sometimes occur because the implemented software is not tested with real (massive data) and the unit tests do not show any bottleneck. In the worst-case scenario, problems occur after the software is rolled out and used by a customer with huge data. In order to prevent such situations, it is recommended to test software with real massive data (if applicable).

2.5 CONCLUSION

At the end of this chapter, we can conclude that efforts aimed at fighting against enormous greenhouse gas emissions and limited energy resources concern computer scientists. Here we surveyed different methods that evaluate energy consumption in computer science by examining algorithm complexity, which plays an important role in this regard. Additionally, we also presented different methods of developing energy efficiency software, like the choice of appropriate algorithms needed to solve specific problems in embedded software. Finally, we can say that the programmer has many possibilities to reduce greenhouse gas emissions in the world.

LITERATURE AND REFERENCES

[1] Black, P.E. 2009. *Dictionary of Algorithms and Data Structures.* U.S. National Institute of Standards and Technology. http://xlinux.nist.gov/dads/HTML/bubblesort.html (accessed February 5, 2013).

[2] Montheu, L. Optimizing the start and execution speed of a distributed monitor and control software in an embedded Linux device. Diploma thesis, University of Ulm.

[3] Levon, J. et al. 2004. OProfile Manual. http://oprofile.sourceforge.net/doc/index.html.

[4] Valgrind Developers. 2012. Valgrind User Manual. http://valgrind.org/docs/manual/manual.html.

[5] Weidendorfer, J., and Zenith, F. 2012. *The KCachegrind Handbook.* http://valgrind.org/docs/manual/manual.html.

CHAPTER 3

Computer Aided Sustainable Design

How Designer and CAD Applications Can Reduce Damage of Products and the Environment

DEFINITIONS AND ABBREVIATIONS

BTU: British thermal unit

CADD: Computer aided design and drafting

CAM: Computer aided manufacturing

CNC: Computerized numerical control

GHG: Greenhouse gas

IPCC: Inter-governmental Panel on Climate Change is the leading body for the assessment of climate change, established by the United Nations Environment Program (UNEP) and the World Meteorological Organization (WMO) to provide the world with a clear scientific view on the current state of climate change and related potential environmental and socio-economic consequences.

ISO: International Organization for Standardization. ISO 14040:2006 describes the principles and framework for lifecycle assessment (LCA) including: definition of goal and scope of the LCA, lifecycle inventory analysis (LCI) phase, lifecycle impact assessment (LCIA) phase, lifecycle interpretation phase, reporting and critical review of LCA, limitations of LCA, and the relationship between LCA phases and conditions for use of value choices and optional elements.

MIL-P-: Military procedure

MTOE: Million tons of oil equivalents

PLM: Product lifecycle management

POTW: Publicly owned treatments works

REDD: Reduced emissions from deforestation and degradation

SIC: Standard industrial classification

STEP: Standard for the exchange of product model data

> **AP:** STEP application protocols
>
> > **AP 203:** Application protocol for configuration of a controlled design. The standard was originally published in 1994 and subsequently amended in 2000 as ISO 10303-203:1994/Amd1:2000(E).
> >
> > **AP 203 E2:** STEP Part 203 Edition 2. Application protocol for configuration of a controlled 3D design of mechanical parts and assemblies. The original AP 203 was restructured as a modular document and published as the technical specification ISO/TS 10303-203:2005(E). After further modification, TS has been replaced by an international standard, published in 2010 as ISO 10303-210:2010(E).
> >
> > **AP 238:** Application protocol for computerized numerical controllers, commonly referred to as STEP-NC. The standard was published in 2007 as ISO 10303-238:2007(E).

TRI: Toxic release inventory

UNFCCC: United Nations Framework Convention on Climate Change

U.S. EIA: U.S. Energy Information Agency

WEEE Directive: Waste Electrical and Electronic Equipment Directive

3.1 INTRODUCTION

Eco-Design reduces adverse impacts of products on the environment, right from the early stage of design and throughout the entire product lifecycle. In this part, we focus on three main points:

- Today's impact of computer aided design (CAD) on Eco-Design processes and how they can be improved in the future.

- How a product designer should organize and plan his or her design.

- Which features should be implemented in CAD and Eco-Design applications in order to simplify a designer's daily work.

3.2 DEFINITION OF CAD

CAD, also referred to as computer aided drafting or computer assisted drafting, is a combination of hardware and software systems that enable architects, engineers, drafters, and artists among others to design and create 2D and 3D virtual models of products with precision. Today, CAD is a relevant system extensively used in the automotive, shipbuilding, and aerospace industries as well as in industrial and architectural design and in many other areas. CAD is also widely used in computer animation to produce special effects in movies, advertising, and technical manuals. Different CAD applications exist on the market, some of which are listed in Table 3.1. There are also many design approaches depending on the designer and company design methods or processes. Figure 3.1 describes the design process of a baseball cap: In steps 1 and 2, different surfaces have been created, trimmed, and joined together, in step 3, material has been added to the joined surfaces, and step 4 shows the final product.

TABLE 3.1   List of Some CAD Applications

Product Name	Company Name
Pro/ENGINEER	Parametric Technology Corporation
CoCREATE	Parametric Technology Corporation
CADDS5i	Parametric Technology Corporation
AutoCAD	Autodesk
Solid Edge	Siemens PLM Solution
TurboCAD	IMSI/Design
CADENAS	CADENAS
PowerShape	Delcam
ArchiCAD	Graphisoft
Modaris	Lectra
CATIA	Dassault Systèmes
Solidworks	Dassault Systèmes
Softplan	Soft Plan System Inc
Cadstar	Zuken

FIGURE 3.1   3D-CAD design steps of a baseball cap.

3.3 HOW CAD AFFECTS TODAY'S ENVIRONMENT

3.3.1 Concurrent Engineering

Concurrent engineering, also known as simultaneous engineering, is a nonlinear product or project design approach in which all phases of manufacturing operate simultaneously. Both product and process design run parallel and occur within the same time frame [1]. Today, using CAD 3D design, people living in different countries can simultaneously work together on the same product or product part. Figure 3.2 shows a cross-functional team with representatives of different departments in different continents. The product manager (from her office in the United States) coordinates a modification review of a part designed by her colleague in South Africa and a supplier in Russia. The sales department is located in Australia and the manufacturer is based in South America. All of these people are also members of the review board and all participants have both write and read access to the CAD server located in Germany and can

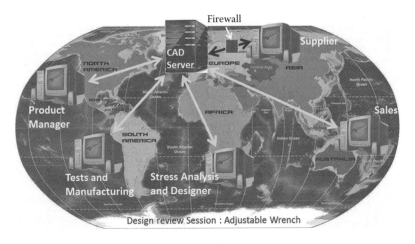

FIGURE 3.2   Concurrent engineering.

model and analyze the same part at the same time. However, the supplier only has restricted access to product data.

Concurrent engineering has benefits. These include reducing product development time, reducing design rework, reducing product development cost, and improving communications. According to the consulting company John Stark Associates [2], companies using concurrent engineering techniques have significant increases in overall product quality, 30 to 40% reduction in project times and cost, and 60 to 80% reduction in design modification after product release.

3.3.1.1 A Practical Example from the Daily Work Routine

In 2010, the average travel cost for a designer going from Hamburg, Germany to New York for a week was estimated at $6000 US (air travel fare: $4000 US; car rental: $100 US/day; lodging: $200 US/day; per diem: $50 US/day). With the introduction of concurrent engineering, the number of designers traveling around the world has reduced significantly and companies are saving money on travel expenses, which can be invested in the research and development of eco-friendly products. According to the aircraft manufacturer Boeing [3], a 747-400 plane that flies 5630 km and carries 56,700 kg of fuel consumes an average of 12 l/km. Based on the "emission calculator" of many airlines, a designer traveling from Western Europe to New York will thus consume approximately 200 l of fuel and will cause CO_2 emission of 480 kg; for a round trip, these amounts are almost doubled. Since the CO_2 emission is considered a factor contributing to global warming, reducing the number of trips made by product designers would have a positive impact on the environment.

3.3.2 Testing of New Products

In order to analyze the level of a product's safety and product systems interactions with their operating environments, the manufacturing company first has to create a product prototype and use it to run certain tests. However, testing can be a highly time-consuming and expensive process. For example, a crash test deliberately turns a brand-new vehicle, which costs thousands upon thousands of dollars, into scraps. However, this does not stop safety engineers from performing hundreds of tests every year. According to the Internet site http://paddocktalk.com, nearly 3000 full-scale crash tests have been performed in Volvo Cars' world-leading crash-test laboratory in just a single decade [4].

Companies invest a lot of money in the prototyping and destroying of products. Ideally, the destroyed products should be collected and recycled. Unfortunately, this environmentally friendly solution is only applied in a few cases. In worst-case scenarios, the destroyed products end up on seashores or as waste in the cities of developing countries [5].

3D CAD enables the creation of virtual models and has a simulation tool that can be used to perform virtual crash tests with almost the same results as physical ones, without destroying brand new models. Thus, using a computer, tests can be performed quickly and inexpensively. This allows for the optimization of design before a physical prototype of the model is manufactured. This means that problems can be solved with simulations before time and money is spent on physical crash tests. Printed output and graphical displays offer great flexibility and give room for designers to solve certain problems that would almost have been impossible to rectify without the help of a computer.

In Figure 3.4, the effect of heat on a grill pan is virtually simulated. On the picture, the designer can identify how the grill pan will react under very high heat. According to the result in the analysis table, he can change some parameters like material, thickness, or geometry of the pan and then rerun the virtual simulation again and again until a satisfactory result is attained.

A physical prototype (Figure 3.3) will then be created when the designer is satisfied with the results of virtual parameters.

3.3.3 The Use of Paper

The negative impacts of paper use on the environment occur at three different levels, beginning with the felling of trees for the retrieval of wood fiber, through to its processing into pulp for making paper, and ending with the disposal of paper products after use [6].

FIGURE 3.3   Prototype of a grill pan.

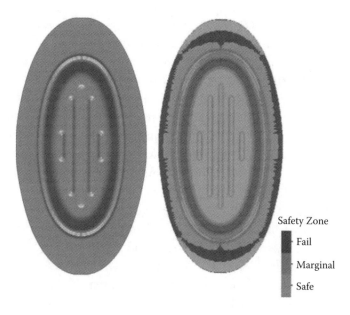

Safety Zone

Fail

Marginal

Safe

FIGURE 3.4   Virtual crash-test—effect of heat on a grill pan.

3.3.3.1 Exploiting Trees for Fiber

According to a report by Campbell et al. [7], forest clearance contributes 20% of the total global emissions of CO_2 into the atmosphere (IPCC, 2007). In addition to harboring as much as 90% of terrestrial biodiversity, tropical forests store more than 320 billion tonnes of carbon. Clearing these forests results in large emissions of CO_2 into the atmosphere; the annual emissions of CO_2 as a result of current tropical deforestation has been estimated to be approximately 1 to 2 gigatonnes or 20% of the total global CO_2 emissions (IPCC, 2007). Reducing forest loss is therefore of utmost importance for mitigating climate change and is reflected in the UNFCCC's commitment to include reduced emissions from deforestation and degradation (REDD) in post-2012 agreements.

The papermaking industry primarily depends on virgin wood-based fibers to make the pulp that becomes sheets of paper. Much of the wood used comes from old growth and environmentally sensitive forests in all parts of the world. In many countries, unsustainable logging not only takes place but also is carried out in areas where it is illegal to log. In many countries, forest destruction additionally displaces native tribes, whose livelihood and cultures are closely tied to the local forest environment.

3.3.3.2 *Processing of Wood Fiber into Pulp for Making Paper*

The pulp and paper industry is very energy intensive, requires enormous amounts of water, and often uses toxic chemicals, the most problematic of which are chlorine compounds used for bleaching pulp to make bright white paper. Although many companies have become more energy efficient, and some even generate part of their power from waste products of wood processing, paper manufacture remains energy intensive and environmentally challenging. Figure 3.5 shows that pulp and paper manufacturers are the fourth largest industrial emitters of greenhouse gases [8].

According to Toxic Release Inventory (TRI) data from SIC codes 261-263, the pulp and paper industry released (into the air, water, or land) and transferred (shipped off-site) a total of approximately 263 million pounds of toxic chemicals during the calendar year of 2000. This represents approximately 2.5% of all TRI chemicals in pounds, released and transferred by all reporting facilities during that year. The pulp and paper industry releases 66% of its total TRI reported poundage into the air; approximately 22% into water and POTW, and 9% is disposed of on land (on-site or off-site). This release profile differs from other TRI industries, which on average dispose approximately 63% on land, 27% into the air, and 4% into water and POTWs. A larger proportion of water releases correlates with water intensive processes of the pulp and paper industry. Air releases contain a variety of toxic chemicals. Approximately 63% is methanol, a by-product of the pulp making process. Other major air polluting

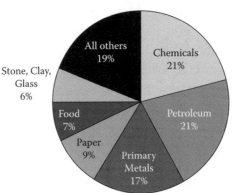

Total Carbon emitted: 371.1 million metric tonnes

FIGURE 3.5 Energy-related carbon emissions for manufacturing industries, 1994. (From US Energy Information Administration, Manufacturing energy consumption survey, 1994. Monthly refinery report for 1994.)

toxic chemicals include chlorinated compounds and sulfuric acid, which originate from the bleaching stage of paper production. Methanol is the most frequently reported chemical disposed by pulp and paper mills and accounts for approximately 15% of water releases and 97% of transfers to POTWs by the industry. Overall, methanol represents roughly 60% of the pulp and paper industry's TRI releases and transfers [9].

3.3.3.3 Disposal of Paper Products

When paper and other organic wastes are deposited into a landfill, they start to undergo aerobic decomposition (in the presence of oxygen). As oxygen diminishes within the first year, small amounts of methane are produced. This is referred to as methanol-genesis. Afterward, when anaerobic conditions take over (in the absence of oxygen), methane production rapidly increases, trapping heat in the atmosphere and creating potentially disastrous greenhouse effects [10]. There is general agreement that methane accounts for up to one-third of human contribution to global warming. Non-recycled paper also contributes to water and land pollution because it is not biodegradable, meaning it is not capable of decay through inorganic or organic means. Thus, when paper waste accumulates, it poses a health threat to people. Poorly disposed waste can also attract household pests, causing urban areas to become dirty, unhealthy, and unsightly places for residence. Moreover, paper not only damages ecosystems, it also reduces available useful land, which would otherwise be on hand for other more useful purposes. The following lines will demonstrate how the CAD can be used to reduce the waste of paper.

By using the 3D CAD modeling, the designer increases his or her flexibility and data integrity. The designer can create or edit parts just by changing the geometry, material parameters, or the relations between design elements (points, lines, planes, etc.) of an existing part. At the end of the design process, the designer can directly forward files of his or her virtual drawings to a CAM application and then to a CNC machine. Only the released version of the drawing is then printed for regulatory agencies.

Figure 3.6 shows how virtual design data is communicated to manufacturing in current practice. "Design" creates the specifications for a product in a 3D model, "Detailing" decides the manufacturing requirements for the product by making a drawing, "Path Planning" generates tools paths, and "Manufacturing" controls production. The job of design is performed using a CAD system and CAD data is subsequently sent to Computer Aided Design and Drafting (CADD) as a STEP AP 203 file. The

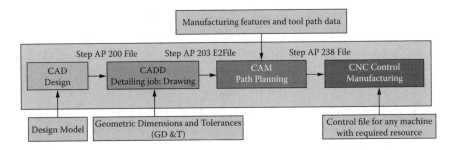

FIGURE 3.6 How design data is communicated to manufacturing in current practice. (From STEP Tools, Inc.)

job of detailing is performed using a drawing CADD system, the CADD file is then sent to the CAM application as a STEP AP 203 E2 file. The job of path planning is performed using a CAM system and the end result is sent to the CNC as a STEP-NC AP 238 file. The job of manufacturing is controlled using a CNC system. In many cases, the CAD, CADD, and CAM functions are combined to a single integrated CAD/CAM system, but the CNC function is always performed by a separate system.

In the past, many designers were forced to mentally retain a lot of information or to write it down on pieces of paper. This frequently led to problems such as component collisions, supply of incorrect quantities, or parts that did not fit. Thanks to 3D CAD, accurate virtual clash detection, precise position check, and exact bill of material are now possible. It minimizes the need for re-work, ameliorates design quality, creates less trash, and reduces the use of paper.

The review manager can send the affected CAD parts and products or a link to the CAD objects by e-mail to other participants for design review, instead of printing the designs and sending them via paper correspondence.

3.3.4 Faster/Better Communication and Customer Satisfaction

With the 3D design, it is possible to communicate and promote interdepartmental understanding earlier in the project cycle. Some designers or manufacturing engineers usually explain their concepts or discuss some issues regarding their product with partners in front of a CAD workstation. They can use the available CAD applications to visualize product status and the different possible configurations. This saves time, minimizes misunderstanding, and creates more room for feedback and control of production operations. The CAD system can help create a common language among groups involved in the development of a product from different backgrounds and jargon. All persons involved in the product development project should at least have reading access to the 3D CAD database. A conversation carried out during a review, a project update, or a sales meeting is always easier and more effective when a 3D model of the product is projected on the screen.

Many companies have now adopted the concept of "environment geometries." The "environment geometry" or "master geometry" is a CAD file created and managed by the department in charge of digital mock up (DMU) and the integration of different parts into a large product. This file is a simple representation of conceptual design of a product containing all of its parts and sub-products in the right position. It is used by engineers to ensure that the parts they have designed fit with the rest of the project. With the master geometry, each department knows exactly where it must place its parts and which department should be negotiated with if more space is needed. This simple and clear space allocation under the product leads to less conflict and misunderstandings between departments and designers.

With the 3D CAD system, the sales departments do not need to wait for the physical prototype before organizing customer presentations, brochures, manufacturing, and technical publications because virtual models can be used.

Thanks to remote connection, the customer and the sales department as well as all departments involved in product development can have a look at and exercise control over the product daily during development phases, even during vacation. Daily check of the product development process helps avoid unpleasant surprises in the end. This means that there will be less need for rework, less energy consumption, less movement of product and engineers between customers and the manufacturing unit, as well as less productions trash.

CAD applications help manufactured products to closely fit the expectations of the customer; this is why many companies have adopted the CAD technology. Jon Peddie Research (JPR) estimated the CAD software market in 2009 to be $5 billion. This is a 23% decrease compared to 2008 when the market was $6.7 billion. The market grew in 2010 but it did not reach the figures seen in 2008 [11].

3.3.5 Enjoying Daylight and Fresh Air

Daylight is good for our health. For example, the rate of depression is lower in rural areas, where people more often go outside in order to work.

Depression and suicide are on the increase, possibly because people experience less daylight [12].

According to EPA, a growing body of scientific evidence indicates that in the last several years, the air within homes and other buildings has become more seriously polluted than the air outside, especially in the largest and most industrialized cities. Other research shows that people spend approximately 90% of their time indoors. Thus, for many the health risk due to exposure to indoor air pollution may even be greater than that of outdoor air pollution [13, 14].

3D CAD applications can be installed and run on Notebooks. In this case, a designer can turn off the lights, desktop computers, air conditioning, and other appliances in his or her office and leave to work outdoors (in a park, for example) if the weather allows. In most cases, it is better to have an upgradable Notebook so that graphic quality and application performance can be improved anytime.

Comparing the 30 W powerful Notebook (including LCD screen) with the 120 W desktop PC and the 80 W CRT screen, up to 80% savings could be achieved. Even in Notebooks intended for use as desktop replacements, with larger screens (up to 16 to 17 in.) and less aggressive power management settings, savings could still be well over 50%. In areas with blackouts and power-surges, the Notebook (batteries included) could save even more. In such areas, desktop PCs would typically require backup from an uninterruptible power supply (UPS) in order to prevent loss of data. A UPS, however, not only generates significant extra cost, it is also a significant energy consumer [14].

3.4 ECO-DESIGN

Eco-Design involves examining the early phase of the product development process for potential environmental damages that a product would cause during its entire lifecycle. In the Eco-Design philosophy, responsibility toward the environment is transferred from the customer to the designer. For this reason, the designer has to think of how long the product should be used and how it should be disposed of or properly recycled. The designer is also responsible for ensuring that raw materials used in product manufacture are renewable and come from either fair trade or eco-responsible businesses. Using Eco-Design, a whole product lifecycle can be simulated with the aim of identifying which stage of the process would cause the largest environmental damages and then looking for more environment-friendly alternatives.

The Eco-Design intends to use resources intelligently and therefore increase benefit for all parties involved through the value creation chain and, at the same time, reduce adverse environmental impact. This should be achieved under socially fair conditions.

The motivation for implementing Eco-Design can be divided into three main aspects, namely:

- Ecological—Human activity damages the biosphere. Exponential population growth, destruction of ecological habitats for the purpose of agriculture and mining, pollution from industry, transportation, and other activities contribute to the damage of the environment. Human alteration of Earth is substantial and growing. Between one-third and one-half of the land surface has been transformed by human action. The carbon dioxide concentration in the atmosphere has increased by nearly 30% since the beginning of the Industrial Revolution. More atmospheric nitrogen is fixed by humanity than by all natural terrestrial sources combined. More than half of all accessible surface fresh water is put to use by humanity and approximately one-quarter of the bird species on Earth have been driven to extinction [15]. To keep the environment intact for future generations, the consumption of non-renewable resources must be reduced as well as activities that have a negative impact on the environment. Eco-Design endeavors to effectively transform these aims into reality.

- Economical—Deploying the concept of Eco-Design in companies can lead to significant economic benefits including saving on materials, reduction in the number of components, use of recycled materials, innovative products with improved functionality, and optimized quality among others. For example, a reduction in the amount of hazardous chemicals in the production processes usually means less effort invested in safety measures (hazardous substances logistics, documentation) and implies cost reduction. In the European electrical and electronic business, WEEE directive imposes the responsibility of disposing electrical waste and used electronic equipment on manufacturers of such equipment. Recycling costs increase according to the quantity of non-eco-friendly manufactured parts. Another very simple example is as follows: Under WEEE, the producer has to pay for recycling according to this equation: Product weight × units sold = market share (in kg). Larger markets share higher recycling costs. If a company is able to reduce a product's weight through eco-design, the calculated market share decreases even without reducing production. It is not yet clear how much recycling will cost under WEEE conditions, but in European countries, where similar systems already exist, the recycling costs per kilogram are usually approximately .50 Euro. Recycling cost would accordingly drop if product weight were reduced [16]. Besides these quantifiable economic figures, Eco-Design can have significant indirect economic effects like better brand image, increased product attractiveness for new markets and consumers groups, advantage in public procurement, etc. In many cases, Eco-Design can lead to reduced material use and energy, which helps save manufacturing and delivery costs.

- Social—Implementing Eco-Design results in socially acceptable conditions, better quality of life, and job creation, all of which enhance social and political stability. Nowadays the idea of Eco-Design is well known across different technical disciplines including architecture, civil engineering, mechanical engineering, and industrial design. Several companies have expressed interest in implementing Eco-Design, because not only laws and regulations oblige them to but also they have realized that Eco-Design is an investment in future-oriented technology, sustainable and responsible business operations, and corporate social responsibility.

Since engineers involved in product development processes today are not necessarily environmental experts, the implementation of Eco-Design still remains a challenging task. Industrial projects reveal that for many engineers, the product lifecycle ends when the product is delivered to the customer. The mentality of "disposal" is still not well established among engineers in the course of product development.

There is still much to be done in the area of product development, considering that many designers still do not acknowledge the influence of a product's design on the environment or, in some cases, the adverse impacts products have on the environment still remain obscure to their manufacturers.

In business today, the Eco-Design is seen as an exotic concept. Many managers and designers perceive it as an obstacle and are more focused on short-term production goals aimed at saving expenditure. They find the Eco-Design concept expensive, time consuming, and without added value in their daily work. Thus, the big challenge for Eco-Design today and in the future is to gain acceptance within the industry.

3.5 DESIGNER IMPACT ON ECO-DESIGN

Significant improvement in eco-friendly product design will only be achieved if designers are truly creative and ready to take product design activities to the next level. An engineer involved in product design needs to understand the product's impacts on the environment. The engineer should also be aware of this fact in order to develop truly sustainable products, designed by taking into account the most environmentally favorable options. The following criteria are hallmarks of a lifecycle mentality.

3.5.1 Do Not Design Products, Design Lifecycles

Instead of designing "green" products, the designer should design product lifecycles. He should think about all the material and energy that go into a product during its entire lifecycle, that is, from "birth" to "death" or better still from "birth" to "rebirth"!

The main activities during the design-planning phase include brainstorming and documentation of facts about material, energy, and toxicity. The importance of these facts is explained next.

3.5.1.1 Material
How will material behave during manufacture, use, and disposal?

3.5.1.1.1 Natural Materials Are Not Always Better   It is easy to believe that "natural" materials are more environmentally friendly than "artificial" or manufactured materials, but is this always true?

Of course, as shown in Table 3.2, the production of wood causes less emission than the production of plastic. However, if one considers the fact that paint and other chemicals are used to preserve the wood and that energy loss is incurred in the process of wood drying and sawing, the answer may be different.

Paint and Other Chemicals Used To Preserve Wood   The U.S. Environmental Protection Agency (EPA) classifies paint as one of the five topmost hazardous substances. Paint generally contains pigment (color), carried by a resin or binder, a solvent, which helps during the paint application process, and a dryer. Vinyl and acrylic paints also include plastic compounds. Others include formaldehyde, arsenic, thinners, and foamers.

Prolonged or increased exposure to paint and paint fumes can cause headaches, trigger allergies/asthmatic reactions, irritate the skin, eyes,

TABLE 3.2   Wood vs. Plastic [17]

Material	Wood	Plastic
Resource base	Renewable	Non-renewable (petrochemical based)
Energy source required for raw material production	Solar	Fossil fuel
Comparable energy consumption for production	1	10
Traps carbon during raw material production	Yes	No
Holds trapped carbon in finished product	Yes	No
Increases total carbon present in global carbon cycle	No	Yes
Off-gassing during and after manufacture	No	Yes
Ability to be salvaged or reused	Yes	No
Percentage of raw material consumed in production	100%	
Biodegradability of post-manufacture and construction waste	Yes	No (photo-degrades, releasing harmful chemicals into watersheds)
Disposability of post-manufacture, construction, and lifecycle waste	Salvage and reuse	Landfill

and air passages, or increase stress on vital organs such as the heart. The World Health Organization (WHO) reported a 20 to 40% increase in the risk of certain types of cancer (especially lung cancer) in people who usually come in contact or work with paint. The National Cancer Institute points out that people who have certain jobs (such as painters, construction workers, and those in the chemical industry) have an increased risk of cancer. Many studies have shown that exposure to asbestos, benzene, benzidine, cadmium, nickel, or vinyl chloride in the workplace can cause cancer [18] and neurological damage.

Energy Needed To Dry Wood   The purpose of drying wood is to reduce its moisture content to a level appropriate for making wood-based products. Most wood-based materials must be dry before further processing. Wood drying involves high-energy consumption and emission levels.

Based on a study we made at SCTB Sarl (*Societe Camerounaise de Transformation du Bois*) a wood processing company in Yaoundé,

Cameroon, approximately 66% of the total energy needed for the entire production of wood-based products is used for the purpose of drying. This proportionality changes according to the desired properties of the product, that is, shape, density, and strength.

Wood Loss through Sawing   During our study mentioned above, we also determined that an important characteristic of sawing machines is that a certain amount of wood fiber is lost during the processing of each piece of lumber. According to engineers working at SCTB Sarl, the lost value of wood fiber is termed *wood loss per sawline.*

Sawmill Improvement Program (SIP) studies are presented in Table 3.3. The purpose of these SIP studies, which began in 1977 and continued until 1988, was to analyze sawing accuracy by machine type.

These studies administered by the U.S. Forest Service show that sawing losses depend on sawing machine type. The band linear resaw, with a wood loss per sawline of 0.206 in., had the lowest significant wood loss per sawline of all investigated machine types.

When it comes to the question of wood or plastic, which option is more eco-friendly? Plastic advocates point out advances in plastics recycling and use of recycled resins to reduce the amount of virgin material needed in some designs. Many plastic product manufacturers claim to recycle all damaged and used products to produce new ones. They also add the fact that deforestation makes significant contribution to carbon dioxide gas emissions.

For an objective answer to this question, we can conclude that while both wood and plastic have individual and unique properties, the most important aspect of "going green" is "getting clean," that is, reducing the amount of products and waste materials that end up on landfills. If manufacturers can find ways of continually reusing products and reducing the

TABLE 3.3   Mean Values of Kerf Width and Wood Loss per Sawline by Machine Type

Machine Type	Machine Code	Number of Machines Studied	Kerf Width (in.)	Wood Loss for Sawline (in.)
Band headrig	1	50	0.162	0.240
Circular headrig	2	168	0.282	0.371
Band linear resaw	3	10	0.139	0.206
Vertical band splitter resaw	4	8	0.158	0.257
Single arbor gang resaw	5	24	0.258	0.311
Double arbor gang resaw	6	6	0.232	0.268

Source: SIP studies.

amount of waste resulting from production with wood or plastic, then the resulting product is ultimately eco-friendly.

Environmentally sound materials do not exist, but environmentally friendly products and services are feasible. The lifecycle mentality helps designers in developing such products.

3.5.1.1.2 Use Minimum Material How little material can be used to achieve the utmost of a product? The concept of using less material may seem obvious, but in fact is more complex than most people would imagine. Very often, the amount of material can be reduced by critically looking at dimensions, required strength, and production techniques. It can even be more beneficial to use materials that have a high environmental load per kilogram, if savings on weight can be achieved. This is because less weight means less fuel consumption during product transportation. According to a U.S. EIA report [19] as indicated in Figures 3.7 and 3.8, approximately 27% of the energy in the United States goes into the transportation of people and goods from one place to another. Automobiles, motorcycles, trucks, and buses traveled over 3.0 trillion miles in the United States in 2007. This distance is almost one-twelfth the distance to the nearest star beyond the solar system. It is like driving to the sun and back 13,440 times.

3.5.1.1.3 Use Recycled Materials In the last decade, designers only focused on making recyclable products with new material. For recycling to

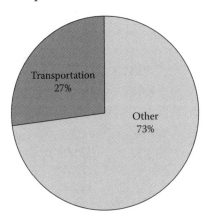

FIGURE 3.7 Quota of total energy used in the United States for transportation in 2009. (From U.S. Energy Information Administration, Annual Energy Review, 2009.)

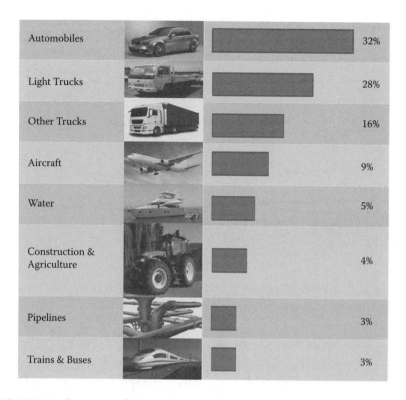

Automobiles		32%
Light Trucks		28%
Other Trucks		16%
Aircraft		9%
Water		5%
Construction & Agriculture		4%
Pipelines		3%
Trains & Buses		3%

FIGURE 3.8   Energy use by various types of vehicles. (From U.S. EIA.)

function, there has to be a market for recycled goods. The more demand there is for recycled materials, the more recycling can be accomplished. People need to recycle their waste and buy products made from recycled waste. If there is a demand for recycled materials, supply will follow suit.

It is good to have recyclable products, but using recycled materials, as much as possible in manufacturing, is even better.

In some countries, like South Africa, the demand for products from recycled materials is growing rapidly [20].

3.5.1.2 Energy

How much energy will a product consume during its manufacture, use, and subsequent disposal?

3.5.1.2.1 Energy Consumption Is Often Underestimated   The environmental impact of energy in the form of electricity or gas is often underestimated as these are not tangible. Figure 3.9 shows the world's primary demand of fuel according to type. Natural gas demand is fast growing,

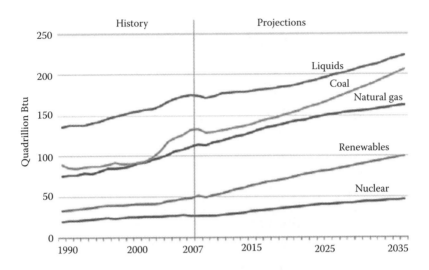

FIGURE 3.9 The world's most marketed fuel by type. (From U.S. Energy Information Administration, international energy outlook, 2010.)

but liquids (e.g., oil) will remain the most demanded fuel in 2030. Energy, as such, is not what the people really need; rather, they need the services that energy can provide like heating, cooling, cooking, lighting, mobility, and motive power.

Thirty years ago, most people drove big cars that consumed a lot of fuel. Fuel shortages in the 1970s did not have much influence on people's driving habits. What changed was the way automobiles were designed and constructed. Auto producers began making smaller and more compact cars with smaller and more energy-efficient engines. In Germany, with the advent of environmental awareness, some household product designs and energy consumption in household heating have continually declined. The Federal Statistical Office, Germany (press release No.372/2010-10-18) reports that the consumption of household energy—adjusted for variations in temperature—dropped by 6.7% between 2005 and 2009.

3.5.1.3 Toxicity

How toxic will a product be during manufacturing, use, and disposal? Here designers should think of the impact of potentially toxic components of chemical, biological, and physical natures.

For example, brominated and chlorinated flame-retardants (BFRs and CFRs) are used in products such as televisions, computers, cell phones, furniture upholstery, mattresses, carpet pads, textiles, airplanes, and cars.

These flame-retardants, also used in a wide range of other consumer products, pose health threats to humans.

Toxic chemicals have been accumulating in the environment and in the human body. Some chemicals are even harmful to unborn fetuses, can affect hormone levels, or even cause cancer.

3.5.2 Increasing Product Lifetime

The lifetime of a product can be influenced in several ways. This includes making the product more durable from a technical point of view or making it updatable (the product can be fixed or easily enhanced) and upgradeable (the old product can be improved to new generations/versions easily).

More importantly, the product should be designed in such a way that people will feel attached to it. Many products are thrown away not because they are damaged, but because their owners easily become bored with them. Today, it is easy to find a person with more than one or two functional cell phones lying around their home not in use.

3.5.3 Do Not Design Products, Design Services

Many customers do not always want a product; they want solutions for problems as well. A service, rather than a product, can be the right solution. For example, joining a carpool or sharing a van to work is a solution for people who need a car only for this purpose. Another example is a bicycle (Figure 3.10) that is designed such that it can be folded into a compact form to facilitate its transportation and storage. When folded, the bicycle can easily be stored in a car, an office, at home, or in a compartment. In this form, it is also ideal for mass transit and commuting or can be conveniently packed in a bike bag and taken along when traveling.

FIGURE 3.10   Bicycle folding process.

3.5.4 Make Recyclable Products

Most products can be recycled; however, only few actually are. Only products that are easily disassembled and have a high enough yield are considered for recycling.

According to the EPA, the United States currently recycles approximately 32% of all waste. This amounts to a savings on greenhouse gases, equivalent of taking 39.6 million cars off the road. Increasing the recycling rate to 35% would reduce greenhouse gas emissions by an additional equivalent of 5.2 million metric tons of carbon dioxide [21].

The designer should increase the chances of making a product recyclable by optimizing its design. To achieve this, a few simple rules should be kept in mind. For example, if thermoplastics are to be recycled, paper stickers should not be put on them and different plastic types should not be combined. For steel parts to be easily recycled, too much copper should not be added into the melt.

3.5.5 Don't Be Afraid to Ask Questions!

Very often, decisions are based on practical experience: "We have always done it this way and it has always worked out well." Very often, big changes in product design and huge improvement on environmental performance, with cost-saving consequences, have been made simply by asking "Why?"

3.5.6 Join an Environmental Organization

Many corporations are members of eco-friendly organizations for brand and marketing strategy purposes, but that is another story. In this section, we focus on the individual. Becoming a member of an environmental organization will be of great benefit to the designer. Such organizations usually provide members a platform of networking with leading professionals in the environmental field and access to innovative, professional, scientific, and technological resources. Additionally, opportunities for professional training and development are possible.

The designer will become more aware of ecological challenges and devoted to the preservation of natural diversity; that is, preserving plant and animal species and their habitats by preventing environmental degradation and destruction. A product designed by a member of an eco-friendly organization is more adapted to environmental conservation.

3.5.7 Customer Education

Customer relationship and marketing strategies are important for any business to flourish. Apart from the sales price, design and appearance are often the two most important parameters that customers take into account when buying a product. The huge challenge for design and sales departments today is to make customers increasingly aware of and consider the impact a product has on the environment before buying it. The customer should be educated on a wide range of environmental topics, like the preservation of natural habitats, organic farming, conservation of endangered species, nutritional treatment for health conditions, renewable energy, reduction of fossil fuel consumption, climate change, animal rescue, and other related environmental topics.

3.6 ECO-FRIENDLY PRODUCT LIFECYCLE

Today, in the face of globalization, simple household products may have traveled over 50,000 km around the world before they are bought at a store. The amount of fuel consumed during the transportation of these products causes considerable damage to the environment. Figure 3.11 shows the amount of fuel used for transportation in the United States in 2009. The percentage of natural gas and biofuels, which are most eco-friendly, only amounted to 3% each.

By applying lifecycle thinking and evaluating products (e.g., by applying lifecycle assessment in accordance with ISO 14040 series), interesting positive results in the environment can be obtained. Thus, the Eco-Design philosophy can be applied to enable a product to cut down its potential negative impact on the environment throughout its lifecycle and at the same time generate economic benefit.

The term "product," according to the online business dictionary http://www.businessdictionary.com, is a commercially distributed good that is tangible personal property, output or result of a fabrication, manufacturing, or production process, that passes through a distribution channel before being consumed or used.

In our case, products also include hardware and software services.

The five lifecycle phases of a product are: (1) extraction of raw materials, (2) manufacturing process, (3) packaging and transport, (4) use, and (5) end of life. All of these are briefly described next.

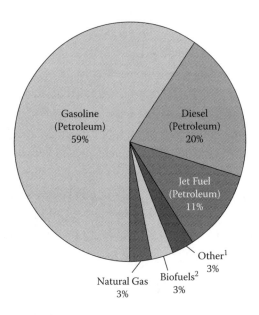

^{1} Electricity, LPG, Lubricants, Residual Fuel Oil
^{2} Ethanol added to gasoline and biodiesel

Note: Due to rounding, data may not sum to exactly 100%.

FIGURE 3.11   Fuel used in the United States for transport in 2009. (From U.S. Energy Information Administration monthly energy review, Tables 2.5 and 3.7c, September 2010.)

3.6.1 Extraction of Raw Materials

In the first lifecycle phase, resources (materials and energy) are extracted from natural raw materials and ancillary materials are produced from the extracted resources. Extracting and processing these constituents consumes natural resources, uses up energy, and causes pollution. The eco-friendly option would be to reduce the material quantity, choose the most appropriate materials, transform waste into raw materials, preferably work with renewable materials, and make products that require only one material type. In this phase, it is also important to verify where the raw material will come from. If possible, use only fair trade materials.

3.6.2 Manufacturing Process

Materials are processed into parts and components during manufacturing. This means components and parts can be assembled to form a product.

In order to manufacture, tools and machines are necessary. The factory, machines, and tools require energy and sometimes create noise. Materials also include chemical substances, coolants, glue, and water.

The eco-friendly option would be to optimize production procedures, methods, and processes to assemble products that are easily separated into their different components during repair or recycling.

3.6.3 Packaging and Transport

Transfer of production factories overseas, cost-cutting, and liberalized markets add up to one thing: The product and its components need to be packaged many times and travel thousands of kilometers before reaching the customer. Depending on the mode of transport (e.g., airplane, ship, train, or truck), a lot of energy is used. At the same time, carbon dioxide is emitted and many noises are generated.

Here, the question is: How little product packaging can be used for the maximum number of customers? The answer is to reduce the amount and volume of packaging to make savings along the chain, from manufacturing to disposal, choose manufacturing sites depending on the customer/consumer locations, use combined transport and alternative fuels, and optimize load.

3.6.4 Use

The product is in use in this phase. To operate properly, the product may require energy or secondary materials, for example, lubricants, water, or coolants. A product has to fulfill its functionality in the use phase or lifetime. The lifetime of the product is the time the product can fulfill its functionality properly. Most customers and consumers come in direct contact with the product in this phase. Any impact on the environment in this phase may also be strain or stress for the user. Here interactions between user and product take place and usually end when the product can no longer be operated properly. The user then disposes it. The product then enters the next phase called the "end-of-life" phase.

3.6.5 End of Life

There are different ways for treating the product and its parts at the end-of-life phase. The product can be disassembled and components reused or the materials can be recycled.

3.7 "GREEN" FAILURE MODES AND EFFECTS ANALYSIS (FMEA)

3.7.1 Definition and History

Failure Modes and Effects Analysis (FMEA) is a task that should generally be performed during the design and development phase of a product's lifecycle. The FMEA discipline was developed by the United States military and formally introduced in production in the late 1940s with Military Procedure MIL-P-1629. Used initially for aerospace/rocket development, FMEA has been helpful in avoiding errors on small sample sizes of costly rocket technology. It was a reliability evaluation technique used to determine the effect of system and equipment failures, whereby failures were classified according to their impact on the success of the mission and personnel/equipment safety.

The primary push for failure prevention came during the 1960s when the technology for placing a man on the moon was being developed. Ford Motor Company introduced FMEA into automotives in the late 1970s for safety and regulatory purposes after the disastrous Pinto affair. Ford Motor Company also used FMEA effectively in production and design improvement. Current advancements in FMEA have come from the automotive sector, where FMEAs are used in all designs and process steps to ensure problem prevention.

Integrated into advanced product quality planning (APQP), FMEA provides primary risk mitigation tools in design and process formats in the prevention strategy. Each possible cause must be examined by a team for its effect on the product or process. This examination should be based on the risk: actions and risk re-evaluation after actions are complete.

FMEA can also be used to assess the environmental impact of a product or process.

"Green" FMEA can identify where corrective and preventive actions can be undertaken to mitigate a product's environmental burden.

It should be performed by a team with designers and environmental and risk experts, who can critically and constructively analyze design and propose tailored solutions.

"Green" FMEA provides a systematic way of answering the following questions:

- How can a product or part fail?

- What will be the impact of this failure on the rest of the process?

- What will be the impact of the failure on the environment?

- How can such a failure be prevented?

3.7.2 Benefits

FMEA is designed to assist engineers in improving the quality and reliability of their design. Properly used, FMEA has several benefits for engineers, the major one being the identification and reduction of product failure, which the customer, the community, and the environment would otherwise experience during the lifecycle of the product.

Among others, FMEA can generate the following benefits:

- Improve quality, reliability, and safety of products, services, machinery, and processes

- Increase customer satisfaction

- Prioritize product/process deficiencies

- Capture engineering/organizational knowledge

- Emphasize problem prevention

- Document and track action taken to reduce risk

- Provide focus for improving testing and development

- Minimize late changes and associated costs

- Catalyst for teamwork and idea exchange between functions

- Improve company image and competitiveness

- Reduce product development time and cost/support integrated product development

- Reduce potential for warranty concerns

- Integrate with design for manufacturing and assembly techniques

- Avoid and reduce cost

- Fewer engineering changes

- Less scrap, rework, and time spent in troubleshooting

- Reduce development cycle time and related cost

- Design optimization

- Can be applied to most products and industries

- Relatively simple principle

- Based on logic and intuition

- Useful to individuals from many disciplines

With FMEA, it is possible to do the right thing right the first time.

3.7.3 Practical Business Scenario

Peter, a designer working for the company Engineers-Pool, is seated in front of his CAD machine, modeling a radiator cap.

While designing, Peter thinks about the sales department requirements, costs, and Eco-design (raw materials, manufacturing, transport, a part's use, end of life, etc.).

When he is done, he saves and checks his part in the database and organizes a review with the FMEA task team.

During the review, a documentation expert must be present. In this phase, good documentation is very important. Each operation in the process should be well described: requirements, failures, effects of failures, and resolutions.

The effectiveness of the FMEA task team depends on the expertise of its members and the quality of the team output depends on the willingness of each team member to give his or her best. The team may include a manufacturing engineer, design engineer, tooling engineer, system safety engineer, industrial engineer, handling specialist, line foreman/operator, customer, materials and process engineer, environment experts, etc.

The agenda of the review meeting Peter organizes should include the following points:

- What can possibly cause a part to fail?

  - Mechanical failure—Will external forces (tension, compression or torsion) be applied to the part?

- Chemical failure—What happens when the part is exposed to chemicals?

- Environmental failure—What is the effect of ultraviolet rays, humidity, microorganisms, ozone, heat, and pollution on the part?

- Is the part recyclable?

- Thermal failure—What happens when the part is exposed to extremely high or low temperatures?

Note that mechanical failure can deform, crack, or break a part into pieces. Thermal failure can warp, twist, melt, or even burn a part. Chemical and environmental failures can change the color of the part or its chemical structure.

- What will be the impact of a failure on the rest of the product? When a part cracks, deforms, or burns inside a product, the product will be damaged and can even hurt the person using it. For example, failure on a radiator cap can cause overheating of an engine. When the engine overheats, it starts to detonate. If the detonation problem persists, damage will occur on the rings, pistons, or rod bearings. The impact of a part's failure on a product can cause:

  - Withdrawal of the product from the market

  - Failure of a product to attain the required market share (its presence in the market is not guaranteed)

  - Inability of a product to achieve the anticipated lifecycle as defined by the organization

- What will be the impact of such a failure on the environment? A failure has environmental consequences if it leads to a violation of any of the known environmental standards or regulations with regard to air pollution, noise pollution, light pollution, soil pollution, radioactive contamination, water pollution, visual pollution, or global warming. Heat, smoke, and noise produced by an overheated motor have a negative impact on the environment.

- Occurrence, severity, and detection of failure cause? To determine this, view failure from the perspective of the customer or prospective user and assess how the other party may experience the consequences. A rating table (Table 3.4) should be created, which evaluates the cause

TABLE 3.4   Rating Table

Occurrence Criteria	Severity Criteria	Detection Criteria	Rating Values
The team is not aware of this failure having ever occurred	Failure would have very little effect	Clear/obvious defect	1
Relatively few failures have occurred	Failures have minor effects	The defect is easily detectable	2–3
Occasional failures, but not in high proportions	Failure causes customer concern	The defect is somewhat more difficult to detect	4–7
Failure expectation	Failure causes severe impact on the end product and customer	Detection may require special inspection techniques	8–9
Failure is almost certain	Failure causes loss of life or a major loss of manufacturing facilities	Defect may elude even the most sophisticated detection technique	10

of a failure, the probability of its occurrence, and detection before the part leaves the manufacturer. The rating table also creates the possibility of failure severity rating, which reflects the seriousness or impact of the failure on other parts of the product or on the customer.

- How to prevent the failure? The risk priority number (RPN) is the product of the occurrence (OCC), severity (SEV), and detection (DET) ratings for each cause of a failure mode ($1000 = 10 \times 10 \times 10$ is the highest value). The decision about the part is made according to the RPN value (Table 3.5). If the RPN value is high because of the severity of effect, a product redesign is likely to be necessary. The FMEA task team can also decide to change some parameters (material, dimensions, etc.) or can decide to stop the design of the part in the company and hire a subcontractor with more experience if the risks are too high.

For the radiator cap, the value of the RPN is 36; it has been made by the formula $OCC \times SEV \times DET$ ($2 \times 9 \times 2$). For this part, relatively few failures will occur and these failures will cause severe damage on the product and customer. The damage will also be easily detectable. After the review, with the results above, the next step for Peter will be to find a way to reduce the effects of the damage (caused by the failures) on the customer and the product.

TABLE 3.5   FMEA Table for the Radiator Cap (Simplified Form of the FMEA Table)

Item	Function	Failure Mode	Local Effect	End Effect	Cause of Failure	OCC	SEV	DET	RPN	Recommended Actions
Radiator cap	Keeps the engine cool by sealing and pressurizing the coolant inside the radiator	Coolant is running out of the radiator	Engine is overheated	Motor damage	Thermal failure: the high pressure of the radiator can deform the cap	2	9	2	36	Change the material or increase the thickness of the cap

3.8 ECO-DESIGN TOOLS

Many software companies are now developing Eco-Design analysis tools as standalone applications or as integrated features in existing CAD or PLM solutions. Like basic stress-analysis and finite-element analysis (FEA) tools, the Eco-Design analysis tools will become standard features of most mid-range and high-end mechanical CAD programs in the near future. The most important thing to keep in mind when developing software is that designers need fast and simplified tools, with simple methods and processes, in their daily work routine.

3.8.1 Standalone Architecture

Standalone architecture describes an Eco-Design application that is not bundled within another software application and that does not require another software package other than the operating system in order to function. It has its own license, serial number, installation directory, and online documentation. It is a more expensive approach, but allows the user better options for customization.

In Figure 3.12, the CAD tool user and the Eco-Design tool user can move data from one system to the other by converting respective files.

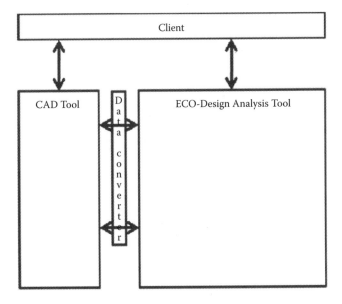

FIGURE 3.12   Standalone architecture.

3.8.2 Integrated Architecture

Integrated architecture describes Eco-Design that operates as features or modules within a CAD software application (Figure 3.13). The best thing about integrated architecture is that data quality and update are consistent and transparent. This is because data conversion or data transfer is not necessary and the results of Eco analysis not only are saved in the CAD file, but also can be printed out as PDF or Excel files.

3.8.3 Vendor and Product Description

There are two big families of Eco-Design applications: software with reliability analysis technology and software with environmental impact analysis technology.

3.8.3.1 Software with Reliability Analysis Technology

These applications offer a suite of comprehensive reliability, availability, maintainability, and safety analysis modules for components and systems

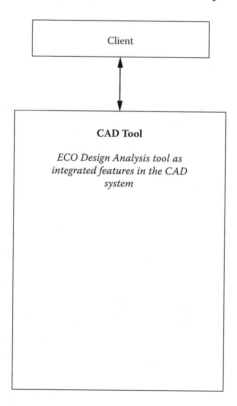

FIGURE 3.13   Integrated architecture.

during the process of product development manufacture. The goal is to perform the following tasks, based on well-established and recognized standards:

- Reliability analysis and prediction

- System reliability analysis

- Maintainability analysis

- Failure modes and effects analysis

- System dependability analysis

- System safety analysis

- System hazard analysis

- Fault tree analysis

- Probabilistic risk analysis (based on FTA and RBDs)

- Sneak circuit analysis

- Lifecycle cost analysis

Over a dozen vendors are currently in the reliability and maintainability support software business. In general, their products are associated with normal reliability and maintainability tasks: prediction and allocation, fault trees, failure modes, effects and criticality analysis, mission simulation, etc. No single vendor covers all analytical bases yet, but four of them (Relex, ReliaSoft, Item, and Isograph) offer a suite of programs that can share a common project system description.

3.8.3.2 Software with Environmental Impact Analysis Technology

These applications are used for environmental analysis, regulations reports, and studies during the product development and manufacturing phases. They offer solutions that enable organizations to operate in a safe and responsible manner, both locally and globally, that is, maintaining compliance with regulatory agencies, and protecting the environment, employees' health, customers, and communities.

This software family is divided into two subgroups: (1) software for emission trading and (2) software for emission monitoring.

1. **Emissions trading (or cap and trade)** is the act of buying and selling carbon credits or carbon/emission permits. A central authority (usually a governmental body) sets a limit on the amount of a pollutant that can be emitted. This limit is allocated or sold to companies in the form of emissions permits, which give them the right to emit or discharge specific volumes of a specified pollutant. Companies are required to hold a number of permits (or carbon credits) equivalent to their emissions. The total number of permits cannot exceed the limit set by the government; as such, total emission is controlled to a certain level. Companies that need to increase their emission rights must buy permits from those who require fewer permits. The buying and selling of permits is referred to as a trade. In effect, the buyer is paying a charge for polluting, while the seller is being rewarded for having reduced emission. Thus, in theory, those who can reduce emissions will do so cost effectively, achieving pollution reduction at the lowest possible cost for society.

2. **Emissions monitoring.** Emissions management and reporting presents companies with new and complex challenges. Monitoring emissions helps address these challenges by delivering streamlined tracking, managing, and reporting direct and indirect greenhouse gas emissions, corporate social responsibility, feedback, worker health and safety, compliance reporting, and emergency response.

3.8.3.3 Emissions Trading, Monitoring, and Compliance Vendors List
On the Internet, there are lists of companies that provide software used in emission trading, emission monitoring, and in the area of compliance. Because of copyright issues, unfortunately we cannot mention some. Many manufacturing companies have also developed internal tools for their needs, and many small providers offer solutions in this area.

3.9 CONCLUSION

In the previous sections, we learned how to design eco-friendly products; we also learned about the immense impact that CAD has on the environment. If CAD applications are mature products, then environmental features or environmental software for GHG reporting is still in a state of infancy. The Eco-Design database of many companies is still on Excel spreadsheets, and even when an eco-analysis tool is available it is used only by a very small group of people because it is expensive, time consuming, and enterprises lack mandatory requirements for reporting GHG emissions.

Environment friendly thinking should be more present during the product development phase. Designer and environmental specialist should work together closely in order to bring better products into the market. There are companies with great expertise in this area, like PRé Consultants (www.pre.nl) in the Netherlands.

Next, we show how CAD software could be adapted to work in a more eco-friendly way in the future.

3.9.1 Eco-Friendly Checker

The eco-friendly checker should be an integrated module or feature operating in the CAD-PLM environment between the design and the manufacturing world.

A CAD model should be released for production/manufacturing if and only if it went successfully through an eco-friendly check process. Each eco-friendly checker should be configured with environmental standards applicable in the customer's country.

3.9.2 Editable Mockups

The CAD software needs to start working more and more with light data. Today, many CAD tools are using the simplified model (Mockup) only for visualization or demonstration; the full representation of the model is only loaded for heavy changes on sketchers or for removing some features.

Simple changes like parameter values, colors, relations, etc. should be performed on the Mockup (simplified representation). This will lead to a reduction in the calculation process and reduce the amount of energy consumed.

3.9.3 Eco-Design Analysis Integrated Features Should Be More User-Friendly

In many companies, it is usually difficult to find someone who knows how to effectively use the Eco-Design analysis features of the 3D CAD application. Most often, features are available but only one or two experts know how to use them. A good Eco-Design feature should be transparent from end-to-end, highly automated, and scalable. It should be easily customizable and adaptable to changing regulations and standards. It should also have the following qualities:

- As many attributes as practically possible with values that are the same as in the real world.

- Different system calculations (the attribute values would not be the same during winter and summer or if the product is to be used in a tropical or cold country).

- Enable the creation of some automatic script and background calculation to reduce the process of data collection and calculation as well as the simulation time.

- The Eco-Design database files (localized default weather data) should easily migrate from one CAD system to another.

- Eco-Design database files should be editable to match a target site's microclimate and should have wide analysis options (energy consumption, air pollution, water pollution, carbon emission, lifecycle cost, etc.).

3.9.4 Define a Clear Added Value Platform for Managers and Engineers

The only way for Eco-Design to definitively enter through the main door of corporations is to make it attractive for managers, designers, and manufacturing engineers. Proof of the fact that lifecycle cost of an Eco-Design product is lower than that of other products should be presented. Government and customer protection organizations should start setting new regulations like value added tax reduction for Eco-Design products.

REFERENCES

[1] Rockford Consulting Group, Ltd. 1999. Concurrent engineering. http://rockfordconsulting.com/concurrent-engineering.htm (accessed October 20, 2010).
[2] John Stark Associates. 1998. A few words about concurrent engineering. http://www.johnstark.com/fwcce.html (accessed October 20, 2010).
[3] Boeing Company. 747 Family: Technical Characteristics–Boeing 747-400. http://www.boeing.com/commercial/747family/pf/pf_400_prod.html (accessed October 20, 2010).
[4] Sulka, M. 2010. Volvo Cars' world-leading crash-test laboratory turns 10. http://paddocktalk.com/news/html/story-135422.html (accessed October 20, 2010).
[5] BBC News. 2010. Trafigura found guilty of exporting toxic waste. http://www.bbc.co.uk/news/world-africa-10735255 (accessed October 20, 2010).
[6] INFORM, Inc. 2008. The secret life series: Paper. http://www.secret-life.org/paper/paper_environment.php (accessed October 20, 2010).

[7] Campbell, A., Kapos, V., Lysenko, I., Scharlemann, J., Dickson, B., Gibbs, H., Hansen, M., and Miles, L. 2008. Carbon emissions from forest loss in protected areas. United Nations Environments Programme—World Conservation Monitoring Centre. A report commissioned by The Nature Conservancy as part of the PACT 2020 Innovation Initiative in collaboration with UNEP-WCMC and the IUCN World Commission on Protected Areas 2008. http://www.unep-wcmc.org/carbon-emissions-from-forest-loss-in-protected-areas_200.html (accessed December 3, 2013).

[8] US Energy Information Agency. 2000. Energy-related carbon emissions in manufacturing. http://www.eia.doe.gov/emeu/efficiency/carbon_emissions/carbon_mfg.html (accessed October 20, 2010).

[9] US Environmental Protection Agency. 2002. Pulp and Paper Industry. Chemical Releases and Transfers. November 2002. http://www.epa.gov/compliance/resources/publications/assistance/sectors/notebooks/pulp-pasnp2.pdf (accessed October 20, 2010).

[10] Biolithe, LLC. All about methane. http://www.biolithe.com/methane.html (accessed October 20, 2010).

[11] Jon Peddie Research (JPR). 2010. CAD Report 2010. http://www.jonpeddie.com/publications/cad_report/(accessed October 20, 2010).

[12] Kripke, D.F. Depression and the droopy dims. *Brighten Your Life.* http://www.brightenyourlife.info/all.html (accessed October 20, 2010).

[13] U.S. Environmental Protection Agency. The Inside Story: A Guide to Indoor Air Quality. http://www.epa.gov/iaq/pubs/insidest.html#Intro1 (accessed October 20, 2010).

[14] European Union ENERGY STAR. Desktop vs. Laptop. http://www.eu-energystar.org/en/en_022p.shtml (accessed October 20, 2010).

[15] Vitousek, P.M., Mooney, H.A., Lubchenco, J., and Melillo, J.M. 1997. Human domination of earth's ecosystems. *Science.* 277(5325): 494–499. http://www.sciencemag.org/content/277/5325/494.abstract (accessed January 26, 2013).

[16] Green EcoSystems Group. 2009. EcoDesign FAQs. http://www.green-eco-systems.com/categoryblog/15-ecodesign.html (accessed October 20, 2010).

[17] Timber Holdings Intl. Woods vs. Plastics. http://www.ironwoods.com/iron-woods_woods_vs_plastics.html (accessed October 25, 2010).

[18] National Cancer Institute. 2006. Risk factors. http://www.cancer.gov/cancer-topics/wyntk/cancer/page3 (accessed December 26, 2012).

[19] U.S. EIA. Using and saving energy for transportation. http://www.eia.doe.gov/kids/energy.cfm?page=us_energy_transportation-basics (accessed October 27, 2010).

[20] Lazenby, H. 2010. Demand for products from recycled materials. http://www.engineeringnews.co.za/article/demand-for-recycled-plastic-timber-recoil-2010-04-02 (accessed October 27, 2010).

[21] U.S. Environmental Protection Agency. Measuring Greenhouse Gas Emissions from Waste. http://www.epa.gov/climatechange/wycd/waste/measureghg.html (accessed October 29, 2010).

CHAPTER 4

Computing Noise Pollution

4.1 INTRODUCTION

The word noise comes from the Latin word "nauseas," meaning seasickness. The origin itself tells us that the word is not something very appealing or enjoyable. The word noise has different meanings depending on the context or subject. In physics, a noise is a random disturbance in an electric circuit that interferes with the reception of a signal. In biology, noise can describe the variability of a measurement around the mean; for example, transcriptional noise describes the variability in gene activity between cells in a population. In computer sciences, sometimes noises are defined as unwanted or meaningless data intermixed with relevant information in the output from a computer.

In the following sections, we will focus on the kind of noise that is an audible, unwanted, distracting, harmful, irritating, or unhindered sound. The fact that noise is invisible, non-palpable, odorless, and tasteless may be the reason why it has not received as much attention as other types of pollution, such as air pollution or water pollution. Whether we are on the street, in water, or in the air, we always hear something. It seems the entire universe is constantly flooded with sounds. For many people, persistent and escalating sources of sound are often considered annoying. Next, we differentiate between sound and noise, and examine when or why sound becomes noise.

It has become increasingly clear that personal computer noise emissions often negatively affect users' well-being and these negative effects undermine diligence of students and employees. Here we outline the basics of noise pollution caused by using computers and the computing systems at home, at school, at a datacenter, in the park, or at the workplace by proposing answers to the following questions: What is computing system noise? Why is my computer noisy? What can be done about computing noise?

4.2 WHAT IS SOUND?

According to the American Heritage Dictionary of the English Language, "sound" is defined as "a travelling wave which is an oscillation of pressure transmitted through a solid, liquid, or gas, composed of frequencies within the range of hearing and of a level sufficiently strong to be heard, or the sensation stimulated in organs of hearing by such vibrations." From the standpoint of physics, sound is an acoustic wave that results when a vibrating source (such as human vocal cords) disturbs an elastic medium (such as air).

We perceive sound because our organs of hearing are sensitive to pressure waves. Perhaps the easiest type of sound wave to apprehend is a short, sudden occurrence like a stone landing in water. When you drop a stone into a river or into a bucket of water, the air around the water is pushed aside as soon as the stone touches the surface of the water. This increases the air pressure in the space closest to the water because more air molecules are temporarily compressed into less space. The high pressure pushes the air molecules outward in all directions at a very high speed (speed of sound). This pressure wave passes through air into space and to the objects and surfaces it touches. That is how the eardrum and the skin receive the pressure wave. When the pressure wave reaches the ear, it cannot be referred to as a sound yet in the true sense of the word, but it is just a vibrating breeze. This is the perception phase as represented in Figure 4.1.

The brain is the only organ that can turn this vibrating breeze into a sound: the processing phase is illustrated in Figure 4.2.

In Figures 4.1 and 4.2, we see that air vibration released from a source can only turn into sound after it is perceived by the ear and processed by the brain.

4.3 HOW CAN SOUND BE DESCRIBED?

In the previous section, we learned that sound is produced when something vibrates. To describe sound in this section, we will use the strings of a piano as a source of vibration. Each piano key is connected to a hammer covered with felt. When a piano key is pressed as in Figure 4.3, the hammer flies up and strikes the strings tuned to generate the corresponding note, then falls away from the string quickly so as not to stop its vibration.

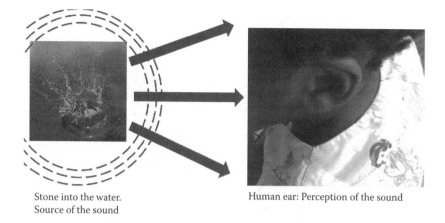

Stone into the water.
Source of the sound

Human ear: Perception of the sound

FIGURE 4.1   Perception phase of sound.

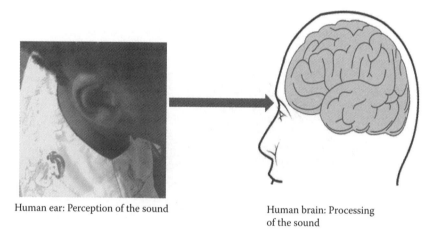

Human ear: Perception of the sound

Human brain: Processing
of the sound

FIGURE 4.2   Processing phase of sound.

Figure 4.4 shows the location of hammers and strings in a piano. Every note sounded on a piano is the result of a string, or set of two or three strings, vibrating at a specific frequency determined by the length, diameter, tension, and density of the wire. When the string of a piano pushes on air, it compresses the air. The vibrating string produces a series of pressure waves. The waves travel to the ear causing the eardrum to vibrate. However, note that you do not hear the strings vibrate, nor the sound waves as they move through the air. It was only after the sound waves have moved your eardrum that activity reaches your cerebral cortex and you can perceive the sound in question.

FIGURE 4.3   A child is pressing the keys of a piano.

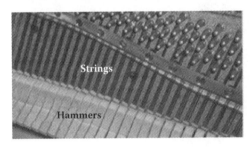

FIGURE 4.4   Strings and hammers in a piano.

4.3.1 Characteristics of Sound

Four characteristics are used to describe the sound wave. These are amplitude, wavelength, period, and frequency and they are represented in Figure 4.5.

4.3.1.1 Amplitude

As the pressure in a sound wave varies, it goes above and below the average pressure. Amplitude can be described as the maximum variation above or below the pressure mid-point. The amplitude of a sound is closely related to its loudness or volume. This can be better understood by studying Figures 4.6 and 4.7.

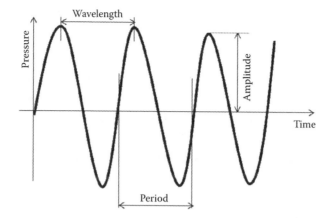

FIGURE 4.5 Characteristics of a sound wave.

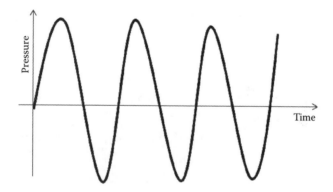

FIGURE 4.6 Loud sound.

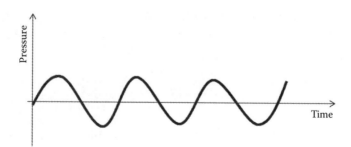

FIGURE 4.7 Soft sound.

4.3.1.2 Wavelength

The wavelength is the distance traveled by the pressure wave during one complete cycle. If you trace your finger across the wave in Figure 4.5, you would notice that your finger repeats its path. A wave is thus a repeating pattern. It reproduces itself in a periodic and regular fashion over both time and space and the length of one such spatial repetition is the wavelength also known as a *wave cycle*. The wavelength is measured by the distance from crest to crest or trough to trough. In fact, the wavelength of a wave is the distance from a point on a wave to the corresponding point of the next cycle of the wave.

4.3.1.3 Period

A period refers to the time it takes to do something. When an event occurs repeatedly, then we say that the event is periodic in reference to the time it takes for the event to repeat itself, which is a period. In the case of a wave, a period is the time needed for a particle (on a medium) to make one complete vibration cycle. Therefore, a period is the amount of time it takes a wave to travel a distance of one wavelength. Period, being time, is measured in seconds, hours, days, or years. The period of orbit of the earth around the sun is approximately 365 days, that is, it takes 365 days for the earth to complete a cycle. The period for the minute hand on a clock is 3600 seconds (60 minutes), meaning it takes the minute hand of the clock 3600 seconds to complete one cycle around the clock.

4.3.1.4 Frequency

The frequency of a wave is defined as the number of complete back-and-forth vibrations of a particle of the medium per unit of time. If a particle of air undergoes 100 longitudinal vibrations in 2 seconds, then the frequency of the wave would be 50 vibrations per second. A commonly used unit for frequency is the Hertz (abbreviated Hz), where

$$1 \text{ Hertz} = 1 \text{ vibration/second} \tag{4.1}$$

The human perception of frequency is called pitch. Very often Hertz and pitch are used interchangeably, but they are actually distinct concepts. Musicians normally refer to the pitch of a signal rather than its frequency.

4.3.2 Loud and Soft Sounds

Sounds are different depending on how loud or soft they are. A loud sound is produced when the hammer in a piano strikes a string with a great force causing a large vibration of the string. This large vibration produces large waves of compressed air. Soft sounds are produced by small vibrations. The more energy the sound wave has, the louder the sound seems as represented in Figures 4.6 and 4.7.

4.3.3 High Frequency and Low Frequency Sounds

A high frequency sound (Figure 4.8) is produced when a string vibrates rapidly. This generates a closely separated series of air pressure waves. On the other hand, low frequency sounds (Figure 4.9) are produced when a string vibrates slowly. The normal range of frequencies audible to humans is 20 to 20,000 Hz (the number of vibrations per second). Any sound with

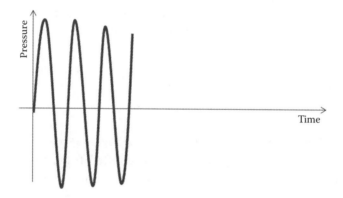

FIGURE 4.8   High frequency sound.

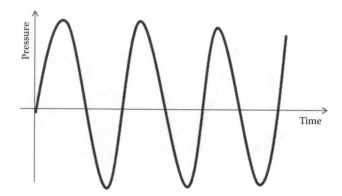

FIGURE 4.9   Low frequency sound.

a frequency below the audible range of hearing (i.e., less than 20 Hz) is known as infrasound, and any sound with a frequency above the audible range of hearing (i.e., more than 20,000 Hz) is known as ultrasound. A range of 200 to 2000 Hz is required to understand a speech.

4.3.4 Single-Dying and Periodical Sounds

4.3.4.1 Single-Dying Sound

A stone landing in a river is a short event that causes a single pressure wave that quickly dies out. Figure 4.10 shows the waveform for a typical handclap. The initial high pressure is followed by low pressure, but the oscillation quickly dies out.

4.3.4.2 Periodical Sound

The other common type of sound wave is the periodic wave. When you ring a bell, after the initial strike the sound comes from the vibration of the bell. While the bell is still ringing, it vibrates at a particular frequency, depending on the size and shape of the bell, causing the surrounding air to vibrate with the same frequency. This causes pressure waves of air to travel outward from the bell, again at the speed of sound. Pressure waves from continuous vibration look more like Figure 4.11.

4.3.5 Formula and Parameters of Sound

4.3.5.1 Sound Pressure, Sound Intensity, and Sound Power

Between sound that is heard and sound emitted from a source, there is a difference due to the distance from the source and the acoustic

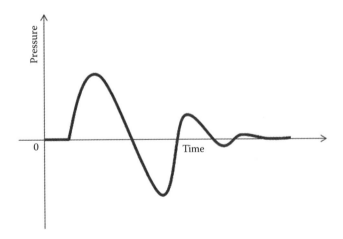

FIGURE 4.10 Single-dying sound.

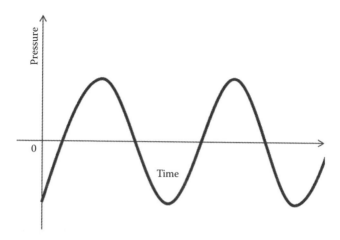

FIGURE 4.11 Periodical sound.

environment (or sound field), in which sound waves are present. To make this difference clear, the physicists of sound have created two expressions: sound power, which is the sound at the source, and sound pressure, which is the sound that we hear. We can simply express this as follows: What we hear or measure with a microphone is the sound pressure caused by sound power emitted from the source.

Sound pressure or acoustic pressure can thus be defined as the local pressure deviation from the ambient (average or equilibrium) atmospheric pressure, caused by a sound wave. Sound pressure can be measured using a microphone in air and a hydrophone in water. The IS unit for sound pressure p is the Pascal (symbol: Pa). Figure 4.12 illustrates sound pressure where 1 is silent sound, 2 is audible sound, 3 is atmospheric pressure, and 4 is sound pressure.

The amplitude of pressure changes can be described by the maximum pressure amplitude, P_M, or the root mean square (RMS) amplitude, P_{rms}, and is expressed in Pascal (Pa). Root mean square means that the instantaneous sound pressures (which can be positive or negative) are squared averaged and the square root of the average is taken [1].

$$P_{rms} = 0.707 \, P_M \qquad (4.2)$$

Sound intensity is a vector quantity determined as the product of sound pressure and the component of particle velocity in the direction of the intensity vector. It is a measure of the rate at which work is done on a

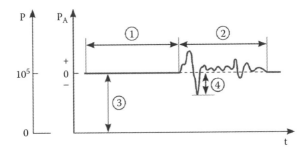

FIGURE 4.12 Sound pressure diagram.

conducting medium by an advancing sound wave and thus the rate of power transmission through a surface normal to the intensity vector. It is expressed as watts per square meter (W/m²).

In a free-field environment, that is, with no reflected sound waves and well away from any sound sources, the sound intensity is related to the root mean square (RMS) acoustic pressure as follows [1]:

$$I = P^2_{rms}/(\rho.c) \tag{4.3}$$

where ρ is the density of air (kg/m³) and c is the speed of sound (m/sec). The quantity $(\rho.c)$ is called the "acoustic impedance" and is equal to 414 Ns/m³ at 20°C and one atmosphere. This acoustic impedance depends on both the characteristics of the wave and the transmission medium. At higher altitudes, it is considerably smaller.

The total sound energy emitted by a source per unit time is the sound power, W, which is measured in watts. It is defined as the total sound energy radiated by the source in the specified frequency band over a certain time interval divided by the interval. It is obtained by integrating the sound intensity over an imaginary surface surrounding a source. Thus, in general, the power, W, radiated by any acoustic source is [1]

$$W = \int_A I \cdot n \, dA \tag{4.4}$$

The multiplication of I with the unit vector n indicates that it is the intensity component normal to the enclosing surface that is used. Most often, a convenient surface is an encompassing sphere or spherical section, but sometimes other surfaces are chosen, as dictated by the circumstances of the particular case considered. For a sound source producing uniformly spherical waves (or radiating equally in all directions), a spherical surface is most convenient and, in this case, Equation (4.4) leads to the following expression:

$$W = 4.\pi.r^2.I \qquad (4.5)$$

where the magnitude of the acoustic intensity I is measured at a distance r from the source. In this case, the source has been treated as though it radiates uniformly in all directions.

4.3.5.1.1 Sound Pressure Level or Sound Level   Sound pressure level (SPL) or sound level (L_p) is a relative quantity measured in decibels (dB) above a standard reference level. It is the ratio between the actual sound pressure and a fixed reference pressure, and can be defined as the threshold of hearing or 0 decibels. The commonly used "zero" reference for sound pressure in air is 20 µPa RMS, which is usually considered the threshold of human hearing (at 1 kHz).

According to Equation (4.6), the sound pressure level (SPL) or sound level L_p is a logarithmic measure of the effective sound pressure of a sound relative to a reference value. It is measured in decibels (dB) above a standard reference level.

$$L_p = 10 \log_{10} \frac{p_{rms}^2}{p_{ref}^2} = 20 \log \frac{p_{rms}}{p_{ref}} = 20 \log_{10} p_{rms} - 20 \log_{10} p_{ref} \qquad (4.6)$$

where p_{ref} is the reference sound pressure (RSP) and p_{rms} is the rms sound pressure being measured [1].

The unit dB (SPL) is sometimes abbreviated just as "dB," which can give the erroneous impression that a dB is an absolute unit by itself.

The commonly used reference sound pressure in air is $p_{ref} = 20$ µPa (rms), which is usually considered the threshold of human hearing. Most sound level measurements will be made relative to this level, meaning 1 Pa will equal SPL of 94 dB. In other media, such as underwater, a reference level of 1 µPa is more often used [1]. These references are defined in ANSI S1.1-1994 [2].

Given a fixed sound source, SPL may vary depending on frequency and diminish with distance. The measured SPL varies with the angle, position, and acoustical environment. As such, single SPL measurements are not a reliable indicator of the overall sound emitted by a source. An SPL measurement is like a single snapshot. It does not show the whole acoustic picture. In air, sound is transmitted by pressure variations from its source to the surroundings. The sound level decreases as it gets further and further away from its source. While absorption by air is one of the factors attributed to the weakening of a sound during transmission, distance plays a more important role in noise reduction during transmission [3].

The sound pressure level of some common sound sources has been measured and the results are available in Figure 4.13 [4].

4.3.5.1.2 What Are Bels and Decibels? A bel is a primary measuring unit for sound volume, named after Alexander Graham Bell, the inventor of the telephone and an early modern acoustician. A decibel (dB) is derived from the combination of deci and bel, meaning one-tenth of a bel. Both

Table of sound levels L (loudness) and corresponding sound pressure and sound intensity			
Sound Sources Examples with distance	Sound Pressure Level Lp|dB SPL	Sound Pressure p N/m^2 = Pa Sound field quantity	Sound Intensity I W/m^2 Sound energy quantity
Jet aircraft, 50 m away	140	200	100
Threshold of pain	130	63.2	10
Threshold of discomfort	120	20	1
Chainsaw, 1 m distance	110	6.3	0.1
Disco, 1 m from speaker	100	2	0.01
Diesel truck, 10 m away	90	0.63	0.001
Curbside of busy road, 5 m	80	0.2	0.0001
Vacuum cleaner, distance 1 m	70	0.063	0.00001
Conversional speech, 1 m	60	0.02	0.000001
Average home	50	0.0063	0.0000001
Quiet library	40	0.002	0.00000001
Quiet bedroom at night	30	0.00063	0.000000001
Background in TV studio	20	0.0002	0.0000000001
Rustling leaves in a distance	10	0.000063	0.00000000001
Threshold of hearing	0	0.00002	0.000000000001

FIGURE 4.13 Table of sound levels.

bel and decibel can be used to describe SPL or sound power. It has become standard practice in recent years, however, to use bel for sound power and decibel for SPL in order to avoid confusion.

According to the Minnesota Pollution Control Agency [5], addition and subtraction of decibels is often necessary for estimating total sound levels or background sound. Because decibels are measured using a logarithmic scale, conventional linear mathematics cannot be used. The most convenient way to perform simple arithmetic functions involving logarithmic measurements is to use doubling rules. These rules provide an accurate estimate of the effect distance and multiple sources have on measured SPL.

4.3.5.1.2.1 Distance Attenuation Estimations　When the distance is doubled from a line source, the sound level decreases 3 dB.

Example:
If a sound level is 70 dB at 50 m, it will be 67 decibels at 100 m, and 64 decibels at 200 m.

When the distance is doubled from a point source, the sound level decreases 6 dB as illustrated in Figure 4.14.

Example:
If a sound level is 95 dB at 50 m, it will be 89 dB at 100 m, and 83 dB at 200 m.

4.3.5.1.2.2 Addition and Subtraction of Decibel Levels　In many situations pertaining to sound control and monitoring, it is very useful to be able to add and subtract noise levels. A doubling of sound energy yields an

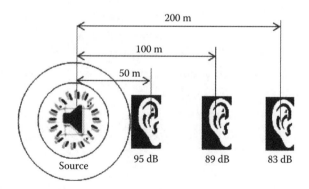

FIGURE 4.14　Distance attenuation of noise levels.

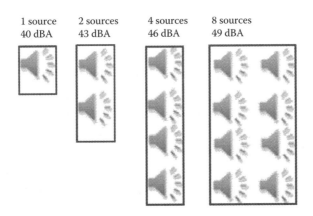

FIGURE 4.15   Addition and subtraction of decibel levels.

increase of 3 dB as represented in Figure 4.15. It is important to note the characteristics of logarithmic addition or subtraction of decibel levels.

Background or ambient noise is present in any environmental noise-monitoring situation.

Background noise is considered all noise sources other than the noise source being monitored. Wind may be a major source of ambient noise. The MPCA's noise test procedures state:

> Measurements must not be in sustained winds or in precipitation which results in a difference of less than ten decibels between the background noise level and the noise source being measured.

4.3.5.1.3 Distance Law   When measuring the sound created by an object, it is important to measure the distance from the object as well because the sound pressure decreases with distance from a point source with a $1/r$ relationship (and not $1/r^2$, like sound intensity).

The distance law for the sound pressure p in 3D is inverse-proportional to the distance r of a punctual sound source [4].

$$p \propto \frac{1}{r} \tag{4.7}$$

If sound pressure p_1, is measured at a distance r_1, one can calculate the sound pressure p_2 at another position r_2,

$$\frac{p_2}{p_1} = \frac{r_1}{r_2}$$

$$p_2 = p_1 \cdot \frac{r_1}{r_2}$$

(4.8)

The sound pressure may vary in direction from the source as well, so measurements at different angles may be necessary, depending on the situation. An obvious example of a source that varies in level in different directions is a megaphone.

4.3.5.2 Acoustic Velocity and Speed of Sound

The speed of sound propagation, c, the frequency, f, and the wavelength, λ, are related by the following equation:

$$c = f\lambda \tag{4.9}$$

A disturbance introduced in some point of a substance will propagate through the substance with a finite velocity. The velocity at which a small disturbance will propagate through the medium is called the acoustic velocity and is related to the change in pressure and density of the substance, and can be expressed as

$$c = (dp/d\rho)^{1/2} \tag{4.10}$$

where
  c = sound velocity (m/s, ft/s)
  dp = change in pressure (Pa, psi)
  $d\rho$ = change in density (kg/m³, lb/ft³)

4.3.5.2.1 Speed of Sound in Gases, Fluids, and Solids The acoustic velocity can alternatively be expressed with Hook's Law as

$$c = (E/\rho)^{1/2} \tag{4.11}$$

where
  E = bulk modulus elasticity (Pa, psi)
  ρ = density (kg/m³, lb/ft³)

4.3.5.2.2 Speed of Sound in Ideal Gases   Since the acoustic disturbance introduced in a point is very small, the heat transfer can be neglected and, for gases, assumed isentropic. For an isentropic process, the ideal gas law can be used and the speed of sound expressed as

$$c = (k\, p/\rho)^{1/2}$$

$$= (k\, R\, T)^{1/2}$$

(4.12)

where
  k = ratio of specific heats (adiabatic index)
  p = pressure (Pa, psi)
  R = gas constant
  T = absolute temperature (°K, °R)

For an ideal gas, the speed of sound is proportional to the square root of the absolute temperature.

Example: Speed of Sound in Air   The speed of sound in air at 0°C and absolute pressure 1 bar can be calculated as

$$c = (1.4\,(287\ \text{J/K kg})\,(273\ \text{K}))^{1/2} = 331.2\ (\text{m/s})$$

where k = 1.4 and R = 287 (J/K kg).
  The speed of sound in air at 20°C and absolute pressure 1 bar can be calculated as

$$c = (1.4\,(287\ \text{J/K kg})\,(293\ \text{K}))^{1/2} = 343.1\ (\text{m/s})$$

Example: Speed of Sound in Water   The speed of sound in water at 0°C can be calculated as

$$c = ((2.06\ 10^9\ \text{N/m}^2)/(999.8\ \text{kg/m}^3))^{1/2} = 1435.4\ (\text{m/s})$$

where $E_v = 2.06\ 10^9$ (N/m^2) and $\rho = 999.8$ (kg/m^3).

4.3.6 Perception of Sound
Auditory organs often include an external structure that gathers, focuses, and perhaps adds resonance to sound. Sound then passes through a tube,

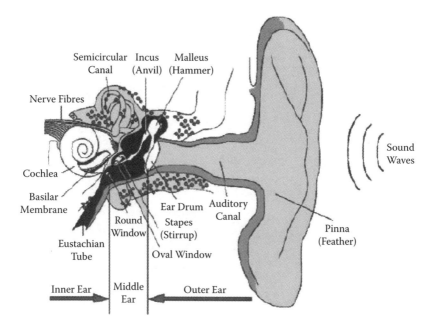

FIGURE 4.16 Sound waves arriving at the human ear.

channel, or chambers, which also can function as resonators. In the following lines, we will use the human ear as an example. Figure 4.16 is a cross-sectional view of the ear showing the major parts involved in hearing. It also shows sound waves reaching the ear [5].

Sound is collected by the funnel-shaped pinna and passed down the air-filled outer ear canal striking the ear drum, which vibrates (similar to the membrane of a microphone). The vibration is conveyed to the air-filled middle ear by bones called ossicles (malleus, incus, and stapes). Finally, vibrations reach the inner ear where the cochlea is located. The cochlea is filled with liquid and contains the hairy sensory cells that convert sounds into nerve signals to be conducted through the auditory portion of the cranial nerve to higher brain centers. This process is illustrated in Figure 4.17 [3].

It is worth noting that humans have less sensitivity at lower frequencies. This is why sound quality appears to diminish when the volume on music is turned down. However, sophisticated music systems augment (amplify) the lower frequency sounds when set to low volumes.

In general, a body part, such as an ear or a membrane, vibrates in accordance with movements of molecules in the surrounding medium (air, water, or soil). This starts the chain of events that leads

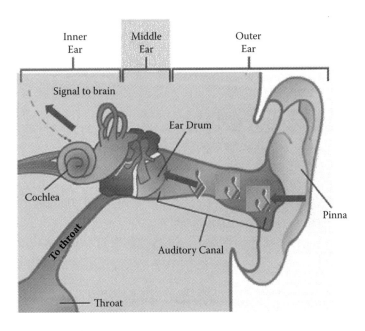

FIGURE 4.17   Sound perception process.

to transduction. While the principles underlying perception of sound are the same for different frequencies, sounds below the normal human range of hearing (infrasound) or above it (ultrasound) sometimes differ in their perception.

The results in Figure 4.18 are classified from pleasant (color green) to unpleasant to the human ears (red)[4]. The threshold of hearing (lowest audible noise level) is approximately 20 μPa. This level corresponds by convention to an SPL of 0 dB. The background noise level in a quiet bedroom at night is approximately 30 dB, while in a typically quiet library it is at approximately 40 dB. A normal conversation heard from a distance of 1 m will be approximately 60 to 70 dB and city traffic is roughly 80 dB. Noise levels above 80 dB over a long period of time degrade hearing, and noise levels of 120 dB and above are perceived by humans as painful.

Thus far, we have established that auditory perception informs us about sounds and their characteristics. Hearing sounds simply involves detecting the frequency and amplitude of these sound waves, the pitch, timbre, and loudness, for example.

Imagine yourselves sitting in the living room watching football, having a good time. Suddenly, you hear loud sounds coming from the neighboring

Table of sound levels L (loudness of noise) with corresponding sound pressure and sound intensity		
Sound sources (noise) Examples with distance	Sound pressure Level L_p dB SPL	Sound pressure p N/m² = Pa Sound field size	Sound intensity I W/m² Sound energy size
Jet aircraft, 50 m away	140	200	100
Threshold of pain	130	63.2	10
Threshold of discomfort	120	20	1
Chainsaw, 1 m distance	110	6.3	0.1
Disco, 1 m from speaker	100	2	0.01
Diesel truck, 10 m away	90	0.63	0.001
Curbside of busy road, 5 m	80	0.2	0.0001
Vacuum cleaner, distance 1 m	70	0.063	0.00001
Conversational speech, 1 m	60	0.02	0.000001
Average home	50	0.0063	0.0000001
Quiet library	40	0.002	0.00000001
Quiet bedroom at night	30	0.00063	0.000000001
Background in TV studio	20	0.0002	0.0000000001
Rustling leaves in the distance	10	0.000063	0.00000000001
Hearing threshold	0	0.00002	0.000000000001

FIGURE 4.18   Table of sound levels.

house. After a couple of minutes, you start complaining to your friends, saying, "This drumming sound is disturbing." So, the question is, how did you make the connection between the sounds you heard and a drum? Sometimes, it is even possible to associate the sounds to a specific drum manufacturer.

Fifteen minutes later, the same sounds keep coming from the neighbor's house. This provokes you to say, "This naughty boy is not supposed to play music at this time of the day. We have to talk to his parents." By hearing these sounds, you have already created a virtual image of a Yamaha drum being hit hard by a naughty boy from next door.

This example shows that sometimes when we hear a sound, we can also determine objects that are the sources of the sound. This means that sound waves travel in space not only with the characteristics of the sound, but also with some information about the source that produced the sound. This information also serves as a basis for the brain interpretation of the sound.

There is a branch of psychophysics called psychoacoustics (Figure 4.19), which is focused on studying the relationship between physical sounds and the brain's interpretation of them.

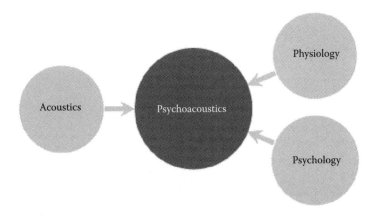

FIGURE 4.19   Components of psychoacoustic.

4.4 NOISE

4.4.1 Definition

If you and your colleagues were jazz lovers, the beautiful and pure voices of Ella Fitzgerald and Billie Holiday filling your office at a moderate volume would be considered a positive atmosphere booster, inspiring, motivating, very pleasant, or promising. At the same time, the repetitive sound of a hydraulic press machine coming from the manufacturing hall, with the same volume, may be considered deeply disturbing. According to this description, we can say that certain types of sound are definitively more bothersome or annoying for some people than others and rather than being described as sound, they are perceived as noise. Noise can then be defined as a loud, unpleasant, unexpected, or undesired sound. According to Environmental Protection UK, noise is the most pervasive environmental pollutant. It has the potential to cause disruption, annoyance, and stress leading to sleep disturbances, interference with communication, and negative health impacts [6].

As mentioned in Section 4.2, noise can also be defined as the brain's interpretation of a physical sound. This interpretation depends on the psychological state of the person at the time he or she perceives the sound waves. This is why sometimes one person's noise is another person's sound. Noise is subjective and can be divided into three categories:

1. Environmental noise, which is defined as the noise coming from the road and determined by the vehicle speed (tires on road surfaces or engine noise, for example), noise from rail or air transport (during takeoff, landing, and turning), and from industry facilities.

2. Local noise is defined as noise affecting a neighborhood or a small area; for example, noise from a football stadium, a discotheque, a shopping mall, or kindergarten. The individual sources of local noise can usually be found easily.

3. Domestic noise disturbs the least number of people, but can be the most adverse and the most complained about noise. Domestic noise is generated in houses, offices, or by planes and cars. The most common domestic noise complaints are about loud music and television volume, shouting, cell phone ring tones, dogs barking, server room noise, and do-it-yourself work.

4.4.2 Perception

Our perception of objects, both visual and auditory, is conditioned by specific principles. These principles function such that our perceptual worlds are organized into the most elementary pattern consistent with the sensory information and with our experience.

Working as CAD-PLM consultants at one of the biggest airplane manufacturers in the world, we interviewed some colleagues in a survey on the perception of noise in our office. We also rented the NTI Audio XL2 (Figure 4.20) in order to measure the sound level and to analyze the acoustics.

Our office was an open office with approximately 5 m width and 16 m length. Compared to enclosed offices, the open office allows for better communication in the team and saves space. Figure 4.21 is a reproduction of our office during the project. We had a meeting table and three islands with two desks each. On the top of each desk, there was a PC, a laptop, a telephone, a small fan, and a desk light. In the room, there was a printer, a heater, and a UNIX workstation.

In the office, we had different sources of noise:

- Machine-produced noise like ringing phones, computer noise, and desk light noise.

- Human-produced noise like phone calls, meetings, and conversation.

FIGURE 4.20   NTI Audio XL2. In the box and the measurement microphone.

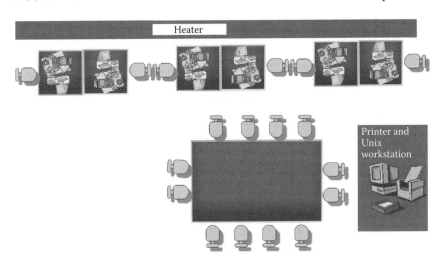

FIGURE 4.21   Inside the office.

- Background noise produced by heating, ventilation, etc.

- Impact noise produced by walking on hard surfaces, opening and closing the doors or the windows, etc.

All these noises were distracting sometimes, but we unanimously agreed that the noise produced by humans was the most distracting because we were able to hear, understand, interpret, and then be involved in the conversation. The noise made by the printer was also annoying sometimes, especially in the morning, but the colleague in charge of printing CAD models and assemblies for collision analyses and Digital Mock-Up review never complained about it.

We concluded our study by assuming that there are variations in individual perception of noises and their intensity. Our perception of noise depends on the interest or relationship we have to the source of the sound. Our study also showed (closely approximated) that sound is perceived to be twice as loud if the sound level increases by 10 dB. Similarly, a 20-dB increase in the sound level is perceived four times as loud by the normal human ear. In hearing, we also tend to organize noises into auditory objects or streams and use the principles of grouping to help us segregate those components we are interested in from others. We are thus able to focus our listening attention on a particular noise source and distinguish an auditory object from the background noise.

Harvey Fletcher and Wilden Munson did research into the perception of the human ear to a wide range of frequencies over a wide range of listening volumes. They created some curves, representing the feedback from subjects who were asked to report when two different frequencies were perceived to be at the same volume. In Figure 4.22, which represents the Fletcher-Munson curves, the threshold of hearing in terms of sound pressure level over frequencies is shown. We can see that the ear does not

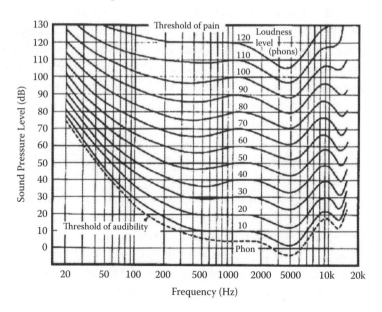

FIGURE 4.22 Fletcher-Munson curves: pain and audibility thresholds for the human ear.

hear all frequencies equally well at different listening levels. For example, to be "as loud as" a 30-dB SPL 1-kHz tone, a 10-kHz tone needs to be approximately 40 dB SPL, and a 100-Hz tone must be more than 45 dB SPL. The threshold of hearing is frequency dependent, and typically shows a minimum (indicating the ear's maximum sensitivity) at frequencies between 1 kHz and 5 kHz. The interesting aspects of these curves are that it is difficult to hear low frequency of soft sounds, and that the ear is extra sensitive between 1 and 6 kHz. The absolute threshold of hearing (ATH) is the lowest of the curves and the minimum amplitude (level or strength) of a pure tone that the average ear with normal hearing can hear in a noiseless environment. The threshold of pain is the SPL beyond which sound becomes unbearable for a human listener. This threshold varies only slightly with frequency.

Prolonged exposure to sound pressure levels in excess of the threshold of pain can cause physical damage, potentially leading to hearing impairment.

4.4.3 Effects of Noise

In recent years, noise pollution has been a topic of heated discussion among researchers because it is often underestimated, despite the fact that it causes a severe impact on human beings and other living creatures, as well as inanimate objects. The adverse effects of environment noise appear in our life in different forms, starting with personal damage via physical and mental impairment and moving to social and economic effects on society.

Some music fans and computer scientists say that intermittent noise is not dangerous, but this is not true; even periodical noise can negatively affect human hearing in the end.

The insidious thing with noise pollution is that, most of the time, the effects do not appear immediately after exposure but come creeping hours or days later.

For example, if somebody hears 5 hours of loud hard rock music in a nightclub, in the morning he will feel that the ambient noises appear dull and further away. However, the actual hearing damage arises only after a longer time. Naturally, hearing damage requires very loud music over prolonged periods, or very loud ambient noises, which accordingly damage the human ear. A nightclub visit does not mean that one inevitably leaves with serious hearing impairment. The number and level of exposures to noises determine the level of impairment.

Next, we summarize some of the adverse effects of noise pollution.

4.4.3.1 Auditory Effects

Acoustic trauma: Sudden hearing damage

Tinnitus: Ringing in the ears

TTS: Temporary threshold shift

PTS: Permanent threshold shift

Interference with communication

4.4.3.2 Non-Auditory Effects

Annoyance

Sleep disturbance

Attention deviation

Motivation reduction

Stress reactions

Cardiovascular problems

Endocrine problems

4.4.3.3 Noise and Sleep

Increased time needed to fall asleep.

Causes awakenings and changes in sleep stages; reduces REM and slow wave sleep.

Sleep disturbance in turn may lead to poor performance and change in mood the next day.

Noise-sensitive individuals and elderly are more vulnerable.

Chronic exposure leads to increase of cortisol.

Adaptation is possible but not complete habituation.

In Europe, 10 to 26% (55 to 143 million) of people have severe difficulties falling asleep or staying asleep.

Chronic sleep disorders lead to loss of efficiency, psychological well being, learning in childhood, social interaction, and driving ability.

Drivers falling asleep cause 20% of accidents on German roads.

4.4.3.4 Noise and Performance

Distracting effects of noise impairs performance especially among children, noise-sensitive, and anxious people.

Type and degree of control is important in degradation of performance.

Learning in schools can be disturbed by high outdoor or indoor noise levels.

Memory: Recall impaired in children.

4.4.3.5 Noise and Hormones

Prolonged exposure to stress leads to immune dysfunction and increased vulnerability to disease.

Environmental stress leads to production of stress-related hormones.

Hypothalamus-pituitary-adrenal (HPA) axis plays a crucial role in this.

Cortisol is closely associated with stress-related health problems.

Noise can affect the HPA function.

Salivary cortisol secretion is prominent during mental work under 90 dB noises and not under quiet conditions.

Circadian decline in cortisol was not observed among workers chronically exposed to noise >85 dB but did reduce with the use of headphones.

4.4.3.6 Noise and Cardiovascular Effects

Acute noise effects: increase in heart rate, blood pressure, and peripheral vasoconstriction.

There is no epidemiological evidence for relationship between noise exposure and blood pressure in adults but this causality has been consistently reported in children.

Increased risk of ischemic heart disease with outdoor levels >65 to 70 dB.

Association has been shown between noise and serum lipids in women and young men.

4.4.3.7 Noise and Mental Health

Community-based studies show high levels of environmental noise associated with depression and anxiety.

Aircraft noise impairs quality of life for children but does not enhance depression or anxiety.

4.4.3.8 Noise and Fatigue

Increased fatigue and irritability is reported after work in noisy environments.

Fatigue and headache are more common among noise-exposed workers in a survey of 50,000 workers.

Reaction time is prolonged after one week among persons exposed to high noise levels, increasing gradually compared to controls.

4.4.3.9 Noise and Children

Aircraft noise: Attention deficit, communication, learning, and memory.

Concentration, motivation, and language acquisition are negatively affected by increased outdoor and indoor noise levels.

Elevated blood pressure is seen with prolonged exposure to traffic noise.

4.5 COMPUTER SYSTEM NOISE

Personal computers have become a major, extensive, and worldly source of noise in modern educational institutions, public institutions, workplaces, and households. Many manufacturers or sellers of IT and telecommunications equipment used to describe their products as "quiet" or "almost silent," without telling us the standard they used to measure the sound power level. The explanation for this is sometimes just simply lack of knowledge, and sometimes an attempt to mislead customers into believing that their product's noise figures are better than they actually are.

There are two ISO standards that have been established as references for the measurement of sound power level for IT equipment. All relevant declarations and information are analyzed by these standards.

ISO 7779—This standard specifies procedures for measuring and reporting the noise emission of IT and telecommunications equipment. The basic emission quantity is the A-weighted sound power level, which may be used for comparing equipment of the same type but from different manufacturers or for comparing different equipment.

ISO 9296—This standard can be used for the noise declaration of PC system units, hard drives, main boards, DVD readers, CD burners, power supplies, printers, projectors, FAX machines, etc. It also specifies the method of determining these values, and specifies acoustical and product information to be given in technical documents supplied to users by the manufacturer and methods for verifying the declared noise emission values given by the manufacturers. The basic values are the declared A-weighted sound power level and the declared A-weighted sound pressure level at the operator or bystander positions.

4.5.1 Measuring PC Noise

To measure PC noise, we decided to perform a comparative study. We chose three different Notebooks, two from the same manufacturer (an old and a new model) and the third from another manufacturer. Due to copyright policy, we cannot publish the manufacturers' names. We will call them PC1, PC2, and PC3. We decided to perform the measurements during six different activities: the PC starting process, idle status, opening a Microsoft Word document process, reading data from the hard drive, playing a DVD, and the PC shutting down process.

Renting an anechoic chamber was too expensive; therefore, we decided to run the test on a quiet night in an isolated house located in a quiet residential neighborhood, far away from bus and tramway stations in Hanover, Germany, with equipment such as air conditioner, refrigerator, wall clock, and gas heater completely turned off. By so doing, we were able to reduce the ambient noise levels down to 25 dB, which is almost the same as in an anechoic chamber.

Behind closed doors and windows, we set a mattress on the floor and a table in the middle of a quiet carpeted chamber, away from the walls and other objects, on a non-resonant table just big enough to hold all the components.

We started the tests and we used a sound level meter (SLM) CEL-620B with a measurement range for sound levels from 20 dB to 140 dB. Because the daily rental rate charged for the CEL-620B was too high (100 Euros per day; 1 Euro = $1.32 U.S. on Thursday, April 5, 2012), we thought it reasonable to stop renting and buy an SLM with less features, but which could provide comparable results. The SLM we bought was a VOLTCRAFT SL-200 shown in Figure 4.23. This sound level detector meets the requirements of EN 60651 (IEC651) and has the accuracy class 2 for general field examinations with the following technical details:

Pick-up time	125/1000 ms
Frequency range	31.5 – 8000 Hz
Accuracy	± 1.5 dB (94 dB/1 kHz)
Measurement range - sound level	30 – 130 dB
Power supply	9 V Block
Dimensions	(W × H × D) 55 × 210 × 32 mm
Weight	230 g
Can be calibrated according to	ISO
Resolution, sound level	0.1 dB

To be able to use it in compliance with standards, the SLM must be calibrated before every measurement using evaluation curve A (dB), that is, it must be checked using an optional sound calibrator and adjusted as required. The following parameters were set in the sound calibrator: 94

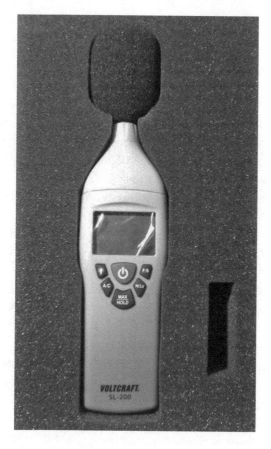

FIGURE 4.23   Sound level meter VOLTCRAFT SL-200.

FIGURE 4.24   Making sound level measurement.

dB at 1 kHz. The curve A (*A-weighting curve*) is used in loudness measurements to accentuate frequencies where the human ear is most sensitive, while attenuating very high and very low frequencies to which the ear is insensitive. The goal is to ensure that value determined by sound level measurement corresponds well with the subjectively perceived loudness.

To measure the sound level of computers, as represented in Figure 4.24, we set the distance between the test objects and the SLM to 1 m from the front, the back, and the sides of the PC according to the most widely accepted standards for noise measurement. We also took a measurement 1 m above the PC. In addition, to ensure that we did not influence the sound waves, the SLM was mounted onto a stand as shown in Figure 4.24.

We turned on the PC and allowed it to start up fully. During the boot process, we measured and saved the highest and the lowest decibel values on the SLM. We also measured when the system was idle. We opened the Microsoft Word application and recorded the noise of the process, then read data from the hard drive and saved the highest value displayed by the SLM. We inserted a DVD into the CD/DVD drive, and repeated the noise measurements. We closed the session and measured the sound level during the shutting down process of the computer.

Figure 4.25 represents PC1. In idle state, PC1 had a noise level of 32 dB, which is very pleasant. Even the moderate noise of 32.8 dB when the Microsoft Word document was opened is still acceptable. However, the notebook was

FIGURE 4.25   PC 1.

challenged during data reading from the hard drive. The sound level of 45 dB measured is quite noisy and corresponds to a noise level between a quiet library and an average home according to Figure 4.18. However, it is true enough that an intensive CPU demanding application will max out the hardware. Playing a DVD movie produced a sound level of 40 dB.

Figure 4.26 represents PC2. In idle state as well as during the opening of a Microsoft Word application, we observed that the value provided by the SLM was constant at 32.2 dB. The sound level can also be described in this case as pleasant. When we started reading data from the hard drive, the value on the SLM increased very fast and reached an average value of 45.8 dB, which is a bit noisy and corresponds to a noise level between a quiet library and an average home according to Figure 4.18. Playing a DVD movie on PC2 produced a sound level of 40.5 dB.

Figure 4.27 represents PC3, which was relatively quiet during the tests, thanks to its low-clocked graphics card and good thermal management. In idle state, PC3 has a noise level of 34 dB, which is very pleasant. Playing a DVD movie produced a sound level of 40.3 dB. A low murmur with

FIGURE 4.26   PC 2.

maximum measure 35.3 dB at full load is outstanding. The keyboard, however, rattled up to 46 dB and as such is not suitable for quiet environments (library).

Table 4.1 represents the measured values in detail.

Comparing the results among each other, as in Figure 4.28, we can say that considering new Notebooks (manufactured in the last 5 years), there is none which in absolute terms is noisier than the others are. It depends on the computing process. PC1 is the quietest during the starting process, but it is also the noisiest during the shutting down process. During the whole test, we did not measure sound levels higher than 50 dB, which is very good. We must also admit that we did not test any computer games or design applications, which are really CPU intensive.

We also compared the results of our tests with the manufacturers' values and the present the results in Figure 4.29. There is a big difference between the manufacturers' values and the measured values for the

FIGURE 4.27   PC 3.

TABLE 4.1   Sound Level Measures of Computing Processes

		Computing Processes					
		Starting Process	Idle	Open Microsoft Word	Shutting Down Process	Data Reading from Hard Drive	Playing DVD Movie
	PC1	33.2–32.1 dB	32 dB	32.8 dB	33.9 dB	45 dB	40 dB
Notebooks	PC2	43–38 dB	32.2 dB	32.2 dB	33 dB	45.8 dB	40.5 dB
	PC3	44–37 dB	34 dB	35 dB	31 dB	35.3 dB	40.3 dB

hard drive accessing process and the idle status. The values are almost comparable for the CD drive accessing.

4.5.2 Why Is My PC Noisy?

To work with a computer, you absolutely need a monitor and a Systems Processing Unit (SPU), which is the box that contains the processor, disk

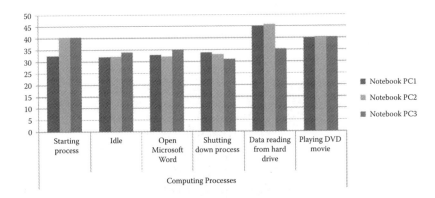

FIGURE 4.28 Sound level comparison chart in decibels.

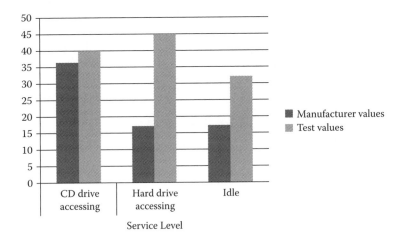

FIGURE 4.29 Comparison between manufacturer values and measured values PC1.

drives, accessory cards, and other electronic components of a computer that are used to process data among others. The SPU is the case that you unplug and walk around with. Within an SPU, the key sources of noise are the following.

4.5.2.1 CD-ROM, DVD, or Other Disc Drive

Normally, the CD/DVD-ROM is quiet when it is not accessed. As soon as the drive is accessed, it starts generating noise. According to the tests we ran, we also figured out that the faster the CD-ROM, the more noise it generates. For example, a 32x CD-ROM drive is far noisier than a 16x CD-ROM drive.

4.5.2.2 Fan

Inside a PC, there are three different types of fans, which are required to cool the system:

The case fans. In many models, these are attached to the front or the back of the PC. It is normal for most additional case fans to produce additional noise. Normally, a case fan is not supposed to make a clicking noise or a high-pitched noise. When this occurs, it is a good indication that the fan is collapsing or is already defective.

CPU fans are not installed on all computers; some computer heat sinks may also contain a fan to help with keeping the processor cool as shown in Figure 4.30. From a CPU fan, you should expect to hear only a minor "snoring" noise. Some issues that may cause additional noise from your CPU fan are a defective CPU fan, CPU fan hitting a cable and causing vibration, and a loose CPU fan causing vibration.

FIGURE 4.30   CPU fan.

The power supply fan is next to where you plug your computer cable to the power supply. Even if today's PCs incorporate additional cooling methods, including auxiliary fans and CPU cooling devices, power supply fans remain the most important elements in the cooling source of a PC. If you notice a lot of noise coming from the power supply area and you are considering changing the fan, first remove dust, dirt, or other elements that contribute to the loud performance of the fan.

4.5.2.3 Hard Drive

Due to the rotation of the drive and the data access, noises are expected from the hard drive. Although hard drive noises are usually nearly negligible, some do make a soft clicking sound when they are being accessed. This is normal.

On the other hand, if you start hearing noises only occasionally or noises that you have never heard before like clicking, grinding, or creaking, your hard drive may be failing. You need to fix or replace it in order not to lose your data.

4.5.2.4 Modem

There are two different types of computer modems: modems with speakers, which are able to emit sound, and modems without speakers—silent modems. The connection noise generated by the modem sometimes helps to check if your computer is still connected to the Internet. Modems can be noisy when they dial up and connect to a remote location. Maybe you work late at night and do not want to wake anyone up or maybe the modem sounds just drive you crazy. Whatever your reason, you can disable (Figure 4.31) or turn off (Figure 4.32) the sounds of your modem by changing your PC settings.

4.5.2.5 Power Supply

The power supply unit is a hardware component inside the computer used to supply the other components with power. Since the power supply unit is an excessively stressed element because of power supply and ventilation of other components, it also generates an excessive amount of noise. If the power supply is failing, the amount of noise coming from the power supply fan will increase.

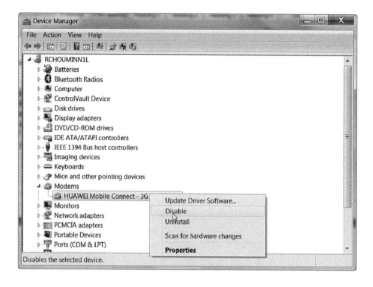

FIGURE 4.31   Disabling a modem in Windows 7.

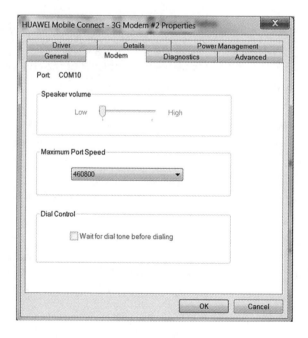

FIGURE 4.32   Switched-off modem Windows 7.

4.5.2.6 Speakers

During the boot or the shutting down process of your PC, it is possible that you sometimes hear some unwanted occasional sound coming from the speakers. At times speakers produce such noise when they are turned off or a device they are connected with is turned off. When this noise occurs frequently, even when the PC is turned on, this is an indication that some components around your speakers are defective or your speakers have started failing.

4.5.2.7 Floppy

The computer floppy drive makes noise during four activities:

1. During the booting or starting up process of the PC, a computer with a floppy drive always verifies if a disk is in the drive and, if this is the case, it will try booting from that disk.

2. When a computer accesses or reads from a floppy drive, it makes a lot of noise.

3. When you run a virus scanner, it also checks the floppy for virus in the drive.

4. During the shutdown process of a computer, the system also accesses the floppy drive in order to check if the disk present in the drive is virus-free. Why? To prevent a save next start of the computer in case this takes place from the diskette.

Most people, when talking about PC noise, refer to noise produced by the SPU. However, other elements also play a contributory role.

4.5.2.8 Monitor

PC monitors are noisy during the following actions: when switched to standby mode, during a resolution switch (e.g., moving from a Microsoft application to a computer game), and when turned off or on. If a monitor produces a lot of noise during a normal operation, this means that it is defective and must be fixed as soon as possible.

4.5.2.9 Keyboard

Keyboards produce noise, but it is not constant and usually not annoying. Sometimes we need the keyboard noise as acoustic feedback to confirm that a key has been pressed. Modern, good-quality keyboards, in general, are better damped than older keyboards and therefore are quieter.

4.5.3 What Can Be Done About Computer Noise?

There are three approaches to countering computer noise: eliminating the source of the noise, using components that produce the least noise, and damping any noise before it reaches your ears.

In this section, we will not study the three approaches separately but we will focus more on the solutions that enable noise-free computing. We will also make a distinction between noise generated by a PC used at home or in the office and noise generated from a data center or server room. We cannot talk about computer system noise without starting at the server rooms and the data centers. A server room is a place where corporations, universities, or other organizations house their computer servers. A data center is a facility used to house computer systems and associated components such as telecommunications and storage systems. It generally includes redundant or backup power supplies, redundant data, communications connections, environmental controls (e.g., air conditioning, fire suppression), and security devices.

Figure 4.33 shows the inside of the server room belonging to a 20-person consulting company.

When entering a server room or a data center, the first impression one gets is that of cold and the noise. It is always a pretty loud, frosty, and stressful experience. Most of the hissing and rushing sounds come from the power supplies and the tools used for cooling. All working servers need to be cooled and the more servers are added to a server room or data center, the more cooling the room will need. Some servers and data center administrators have learned to live with this noise, but others have adapted by wearing soft foam earplugs at their place of work.

Data centers around the world are changing at a rapid pace. Trends such as Cloud Computing, processing of large unstructured data sets (Big Data), and Virtual Desktop Infrastructure are some of the reasons why the demands of modern data centers are growing steadily. According to Michael Vollrath, founder and CEO of Michael Vollrath IT services, which has managed and optimized the services of many data centers in Germany, the current trend in the business is to bring more equipment and denser and hotter systems into data centers. Noise, of course, differs from one center to another or from one system to another. It is now

FIGURE 4.33 Server room.

common for data center workers to spend a lot of time away from the data center location and to manage systems remotely.

Vollrath says that the sound-level measurements in data centers and server-rooms are not priorities in the IT business, but they should be. In the consulting company where we made Figure 4.25, we also measured decibels in 10 different places within the server room. The lowest was 74.9 dB, and the highest was 80 dB, recorded near the heating, ventilation, and air conditioning equipment. At these decibel levels, you have to talk loudly to be heard, but they are considered safe according to the laws in many countries. For example, in 1981 in the United States, the Occupational

Safety & Health Administration (OSHA) implemented new requirements to protect workers in general industries (e.g., the manufacturing and the service sectors) for employers to implement a hearing conservation program where **workers are exposed to a time weighted average noise level of 85 dBA** or higher over an 8-hour work shift. Hearing conservation programs require employers to measure noise levels, provide free annual hearing exams and free hearing protection, and provide training and conduct evaluations of the adequacy of the hearing protectors in use unless changes to tools, equipment, and schedules are made so that they are less noisy and worker exposure to noise is less than the 85 dBA [7].

After talking many times with Michael Vollrath who has been in the data center management business for 20 years, we figured out that in many countries, there is no agreed-upon standard for the decibel level above which data center workers must be protected. However, many IT companies are working on future scenarios in order to protect the data centers and server room workers from noise at their workplace and make their working environment more pleasant. Some of these scenarios have already been successfully deployed. Next, we examine some of them.

4.5.3.1 Data Centers and Server Rooms

4.5.3.1.1 Lights-Out Server Room   A lights-out data center is a server or computer room that is physically or geographically isolated from the rest of a building and the majority of the people working with it. The data center can even be housed in a separate building that may be miles away or even in another country. In this case, the servers contained in the server room or the data centers are under lock and key and kept in the dark. No human workers or administrators need to be inside during normal operation and all operations in the room are automated and controlled by the use of KVM switches to help assure the security of the locked room. KVM is short for keyboard, video, mouse switch—a hardware device that enables a single keyboard, video monitor, and mouse to control more than one computer at a time.

To extend the distance up to several hundred feet between the user and computers, the keyboard, video, and mouse plug into a "KVM extender," which is cabled to its counterpart unit at the computer. Some KVM switches support terminals at both ends of the connection, allowing local and remote access to all of the machines. KVM devices can also control machines across the Internet or attached to a LAN. Software in the client PC converts keyboard, video, and mouse signals into IP packets that are sent over the network. At the receiving end, the IP-based KVM switch is cabled to the computers. With an IP-to-KVM device, which converts IP packets to KVM signals, regular KVM switches can be used in an IP network.

With the lights-out server room, noises generated in the room should not be a big deal for the workers because they are seated somewhere else. If a technician wants to replace a device in a rack or fix something else, he need not stay in the data center for a long time and if so, he should use ear protection devices that reduce workplace noise and noise caused by power tools.

Potential problems with using lights-out data centers include resource management, climate control, troubleshooting, and all other tasks that must be handled remotely.

4.5.3.1.2 Submersion Technology   When electrical devices are exposed to heat sources, they lose their reliability and longevity; that is why in user manuals of all electrical tools, you will always read the note: *Don't place the device near heat sources, like radiators or others.* Engineers all over the world have always worked hard to find solutions to reduce the heat intensity near electrical parts. In computer sciences, air-cooled computing systems have always been the standard.

The idea of cooling some data centers or server rooms' equipment with liquids could seem inconsequential because of rust in particular or corrosion in general. In 1985, Cray-2, the first supercomputer using a liquid cooling process, was developed, mainly for the U.S. Departments of Defense and Energy. Figure 4.34 represents Cray-2 and its Fluorinert-cooling "waterfall." This experience was not a huge success because the liquid used as a coolant was seen as environmentally unfriendly at that time.

In the last five years, the concept of liquid cooling in data centers is celebrating a huge comeback thanks to new green corporate policies and most of all low energy cost, new, safe, odor-free, and high-tech liquids that can absorb heat more efficiently than the older standard liquid-cooling systems.

The submersion technology process is simple. The primary heat-generating components are submerged in the liquid. Most of these new coolants are thermally conductive liquids but do not conduct electricity so they can be in direct contact with electronic components without causing any harm, and because they are many times better than air at capturing and conducting heat they offer the potentiality of dramatically more energy-efficient

FIGURE 4.34   A Cray-2 at the Computer History Museum in California.

cooling than is possible using the standard chilled air approach. The heat captured by the coolant is normally transferred to a liquid pipe loop using a heat exchanger and the heated hot liquid is then pumped out of the data center where the heat can be used to warm office spaces or to provide hot water in a building, further reducing corporate energy costs. In addition, because liquid cooling is approximately soundless, unlike computer room air conditioning and air moving equipment, data center noise levels can be considerably reduced. This makes it easier to comply with the noise regulations in many countries and almost certainly limits the need for staff to use ear protectors when working close to the server hardware.

The industry is now finally recognizing the huge advantages that liquid cooling has to offer. The technological advantage as well as the energy efficiency will no doubt accelerate its acceptance into the server architecture and data center environment.

4.5.3.2 Personal Computer/Standalone Computer

For PCs used at home by gamers or designers, there are also many methods for reducing noise, some of which we explain next. Before moving forward, make sure your computer is clean and dust-free. If you have had it for more than six months, chances are a film of dust has settled on the fans and heat sinks causing it to become hotter (and thus louder), so grab a can of compressed air and clean it out. When you are done, you may find that it is already a little quieter.

4.5.3.2.1 Air Cooling with Larger/Upgraded Fans that Are Less Noisy   Note: The size of the fan should be chosen according to the dimensions of the computer case.

This method is by far the most economical and generally used method for cooling a PC. A large fan is blowing room-temperature air on a cooling device (heat sink), placed over the elements that need to be cooled. The cooling device (heat sink) usually has a flat surface that is in contact with the part to be cooled, and on the other side, several fins are attached. These fins will increase the surface of the device and therefore the heat exchange capability of the device. The fan is blowing air between those fins causing quick and more efficient exchange, rapidly removing the heated air produced between the fins. On the market today, one can already find some fans used for cooling CPUs that are furnished with an integrated metal heat sink that increases the cooling power.

Figure 4.35 represents one of the common processes of CPU air-cooling. A fan is pulling air past the heat sink, and as the air moves past the heat sink it draws the heat away from it. The air in the PC case is drawn out of the case by other fans. As you can see, there is a lot of air circulating around.

Other components such as the fan controller and the fan bearings are also important factors when it comes to decreasing computer noise.

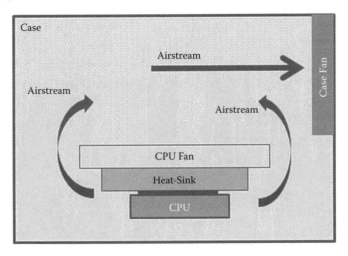

FIGURE 4.35   CPU air cooling process.

4.5.3.2.1.1 Fan Bearing There are many types of fan bearings. Below we describe the most common types.

- Sleeve bearing: This is the most common type. It is cheap and easily available, but also the most noisy. Common sizes (diameters) are 80 mm, 120 mm, and 200 mm.

- Ball bearing: This is a more expensive and quieter version, with the same sizes of sleeve bearing fans.

- Fluid bearing: This is an even more expensive alternative, with even less noise and somewhat rare. It is usually used in high-end HDDs and is the same size as sleeve bearing fans.

- Magnetic bearing: By far the most expensive and exotic version, the fan blades "float" due to magnetic levitation, making it virtually inaudible.

4.5.3.2.1.2 Fan Controller If your fan is running at full speed even if not needed, a fan controller is better to help manage fan speed, so your computer is kept cool when necessary and silent when not. There are many methods to regulate fans (software applications, BIOS settings, external fan controller), many of which are freely available. In the following, the fan controller software application SpeedFan will be downloaded, installed, and run.

The Internet can be used to search for a free download of the software SpeedFan. When found, it can be downloaded to your PC for installation.

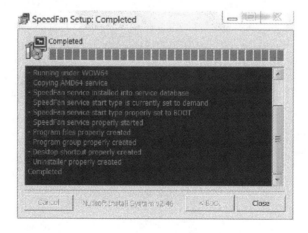

FIGURE 4.36 Completed installation of SpeedFan.

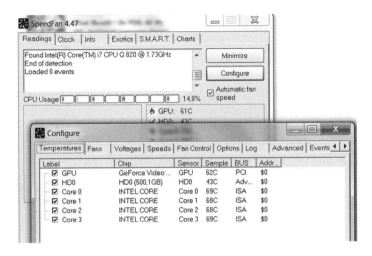

FIGURE 4.37   Information from the SpeedFan.

After completing the installation as represented in Figure 4.36, you can run the application.

Figure 4.37 shows the information delivered by SpeedFan. This gives the most control over the fans by letting you choose how fast they are running at any given time. SpeedFan can even monitor the computer's temperature (see Figure 4.37) and adjust the fan accordingly. By properly configuring SpeedFan, the speed of the fan can be changed based on system temperatures. When choosing parameters for the minimum and maximum fan speed, you should try to set them by hand (disable all the VARIATE FAN checkboxes) and listen to the noise. When there is no noise from the fan, then we can set that value as the minimum fan speed for that fan. It is suggested to use 100 as the maximum value, unless you hear a lot of noise from it, in which case the maximum speed could be reduced to 95 or 90. SpeedFan's configuration can even be used to set the desired temperatures and have it adjust fans automatically based on these temperatures.

Active air cooing is an efficient way of cooling in terms of power saving but has one main drawback: It can only reduce the working temperature of a part to temperatures that are always higher than the environment temperature. This could be a problem when a PC is working in harsh environments or if near the PC there is other equipment that could produce high temperatures when in operation.

4.5.3.2.2 Air-Cooling without Fans   This is also known as passive air-cooling. It does not use a fan or other means of forced air-cooling but relies only on natural convection cooling. Passive cooling has two big advantages: it has no moving parts and operates in silence. However, as components advanced and heat generation increased, the heat-sink surface area required to dissipate additional heat became too large and designers started incorporating small fans onto heat sinks in order to keep their sizes small.

As mentioned before, cooling plates can be very heavy and sometimes require special parts to be fixed over the part to be cooled so that they will not mechanically damage the part or the PCB (printed circuit board) itself.

Passive air-cooling is the most efficient way of cooling in terms of power saving, as it actually needs no power at all to operate. This method has a major drawback: weight! Heavy and large plates must be fixed over small parts, increasing the total weight of a computer and reducing the usable area inside the box. In addition, the ambient temperature should not be very high, as this would make the passive air-cooling inefficient. In many cases, the computer housing has one or two fans to circulate the air inside the computer.

4.5.3.2.3 Water-Based Cooling   The term water-based cooling sounds more like terminology of the automotive industry. Of course, the water-based cooling process has been a part of the automotive engine for more than 60 years; it has been commonly used for cooling automobile internal combustion engines. After reading this small introduction, the following question arises: Why use water-based cooling instead of air-based cooling which seems to be more natural? To answer this question, we must first compare air-cooling to liquid cooling. When comparing the effectiveness of cooling methods, two properties matter most: thermal conductivity and specific heat capacity.

Thermal conductivity is a physical property that describes how well a substance transfers heat. Thermal conductivity, which is material based, will not differ with the dimensions of a material, but it is dependent on temperature, density, and moisture content of the material. The thermal conductivity normally found in tables is the value valid for normal room temperatures. This value will usually range between 273 and 343 K (0 to 70°C). According to the thermal conductivity table, at a room temperature of 293 K (20°C), the thermal conductivity of liquid water is approximately 0.61 W/mK that of air, which is 0.025 W/mK. Obviously, this gives water-based cooling a huge advantage over air-cooling because water-based cooling allows for a much faster transfer of heat.

Specific heat capacity is also an important physical property, which refers to the amount of heat required to change a unit mass of a substance by 1° in temperature. It can be expressed as:

$$c = dQ/m \, dt \tag{4.13}$$

where
dQ = heat supplied (kJ)
m = unit mass (kg)
c = specific heat (kJ/kg °C, kJ/kg °K)
dt = temperature change (K, °C)

According to the specific heat capacity tables, the specific heat capacity of liquid water (4186 J/(kg K)) is approximately four times that of air (1038.13 J/(kg K)), which means it takes four times the same amount of energy to heat water than it does to heat air. Once again, water's ability to absorb much more heat energy without increasing its own temperature is a great advantage over air-cooling.

Water-cooling is seen as a new trend in the PC cooling system. The basic system as represented in Figure 4.38 is made of cooling blocks (also known as water-blocks), that is, tubes within which the coolant flows, a small coolant reservoir, a circulation pump, a radiator, and the water (coolant) that is pumped from the reservoir into the tube that transfers the coolant to where it is needed. Each component to be cooled has a cooling block attached to it. A thermal paste/compound is also applied between each component to be

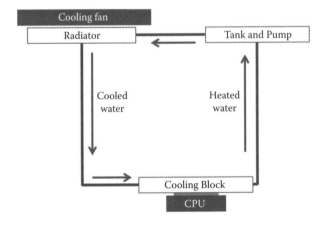

FIGURE 4.38   Water cooling process.

cooled and its cooling block. This block, usually made of copper or aluminum, is an unfilled or hollowed plate with an input and an output for the coolant.

In Figure 4.38, we can see that the water is pumped from the tank to the radiator where it is cooled by a fan. Cooled water comes out of the radiator and is then transported to the cooling block. As the water travels through this cooling block, it transfers heat along with it. This works much more efficiently than air can. The heated coolant is then pumped into a reservoir, from where it travels into a radiator and the cycle begins anew.

Note: This process can run inside or outside the computer case. In our example, it takes place outside the computer case.

4.5.3.2.4 Bong Cooling   As represented in Figure 4.39, the bong cooling process is a combination of air and water-based cooling. The key factor behind the bong cooling is evaporation. The heat generated by the CPU is used to heat water, which is then pumped inside a long tube. The showerhead and the height of the bong allow the water to be separated into smaller droplets that are cooled faster. The cooled water then drops down into the tank at the bottom of the bong, where it is pumped to the cooling block used for cooling the CPU. The output of the cooling block is hot water, which is pumped into the tube and cycled along the loop again.

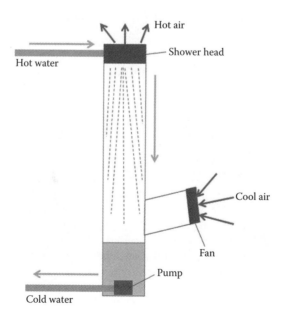

FIGURE 4.39   Bong cooler.

The bong cooling process is not very quiet; there is a consistently smooth and tranquil waterfall sound to be heard during the process. In case this sound is not wanted, it would be better to install a fine-mesh screen filter just above the water level because the waterfall is generated during contact of the smaller droplets with the surface of the water in the reservoir.

Two main issues with the bong cooling are space and the hot humid air that is liberated in the room or the environment.

4.5.3.2.5 Nitrogen-Based Cooling   Nitrogen-based cooling is another emerging technology for quietly overclocking a computer. Unlike conventional approaches, it can be used in a closed-loop or an open-loop system to remove heat. Liquid nitrogen is nitrogen in a liquid state at extremely low temperatures. It is produced industrially by fractional distillation of liquid air. Liquid nitrogen is a colorless clear liquid with density of 0.807 g/mL at its boiling point and a dielectric constant of 1.43. Liquid nitrogen is often referred to by the abbreviations LN2, LIN, or LN.

4.5.3.2.5.1 *Closed-Loop System*   When we started testing this process, it was a bit difficult because many components needed for testing were not easy to get. In the process represented in Figure 4.40, the motherboard is fully submerged in Fluorinert, an electronic liquid offering unique properties, ideally suited for the demanding requirements of electronics

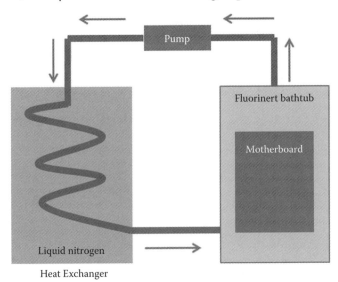

FIGURE 4.40   Closed-loop nitrogen cooling process.

manufacturing, heat transfer, and other specialized applications. It is non-flammable, non-corrosive, has low toxicity, good stability, high dielectric strength, and is compatible with sensitive materials. It is used to immerse devices to test physical and electrical quality. Devices dry quickly without residue to be cleaned. It is manufactured by 3M, a company based in St. Paul, Minnesota.

In the first step, the Fluorinert cools all the components of the motherboard by absorbing the heat generated by them. In Figure 4.40, we see that the heated Fluorinert is pumped out of the bathtub (a JB Packaging polystyrene box) into a heat exchanger where it is cooled by the liquid nitrogen and pumped back into the Fluorinert bathtub. We wanted the heated Fluorinert to be extremely cooled down by the heat exchanger on one hand, and on the other hand, to reach the bathtub with a temperature that will allow startup of the computer without any problems. This is why one of the big challenges of this test is the selection of material for the heat pipe. After some experimenting with copper, aluminum, and nickel, we finally decided to use stainless steel as material for the heat pipe because certain working fluids like nitrogen and Fluorinert need a compatible tube material to prevent corrosion or chemical reaction between the fluids and the tube. Corrosion will damage the tube and a chemical reaction can produce a non-condensable gas.

4.5.3.2.5.2 Open-Loop System   After running the closed-loop nitrogen cooling system, we thought that it could be a simple and cheaper way to quietly overclock a computer with nitrogen liquid. In this process, which we called open-loop nitrogen cooling system, we isolated the motherboard by disconnecting all its cables and removing it from the computer case. When the motherboard had been taken out, we focused only on cooling down the CPU. We removed all the brackets around the CPU making it completely free. All the small components around the CPU were first covered by a clear packing tape and then by modeling clay to avoid corrosion and short circuit. Figure 4.41 shows a nitrogen pot (a protected cylinder can also be used) on the top of the CPU. Before this, thermal compounds were squeezed on the CPU.

Before pouring the liquid nitrogen into the nitrogen top, we made sure that all components were protected and everything was sealed hermetically. In other words, the system should be air tight and waterproof.

After positively checking everything, we carefully carried the motherboard to its working station and reconnected its cables to the computer

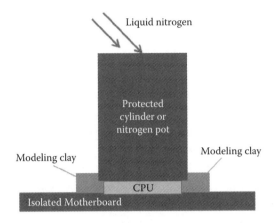

FIGURE 4.41　Air-tight, sealed open-loop nitrogen cooling system.

case. We poured liquid nitrogen into the nitrogen pot and then started the computer. While the computer was running, we poured liquid nitrogen inside the pot again from time to time.

4.5.3.2.6 Acoustic Tiles　Mounting acoustic tiles on walls and replacing ceiling tiles with new acoustic tiles can make a significant difference in the perceived sound, even if the measured decibel level only drops by one or two. There is a lot of performing acoustic tile products on the market today, which are acoustically more efficient and extremely lightweight.

4.5.3.2.7 Active Noise Control　Active noise control (ANC) is a method for reducing unwanted sound. According to Figure 4.42, when a sound is generated from a source, a noise-cancellation speaker emits a sound

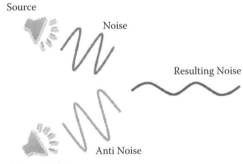

FIGURE 4.42　Active noise control process.

wave with the same amplitude but with inverted phase (also known as antiphase) to the original sound. The waves combine to form a new wave in a process called interference, and effectively cancel each other out. This is called phase cancellation.

In the computer-manufacturing field, this process is still in its infancy. Some researchers and small businesses are coming up with products and solutions that have not been embraced by major computer manufacturers and the community. One of the most popular solutions is to confine the propagation of much of the fan noise to a channel or a pipe, and then use ANC to minimize the strength of the fan noise leaving the channel or the pipe.

4.5.3.2.8 Others   Many measures can be undertaken against computing noise, for example, with smaller cooling units (smaller AC) on each computer heat generator components. We can solve the issue, too. A better connection between all the components involved in the computing process and utilization can help avoid some noise (e.g., the zzzzz noise coming from poorly connected external speakers). A hearing protection sign can also be mandatory in all data centers and server rooms or on the package of some computer games.

REFERENCES

1. Hansen, C.H. 2003. Fundamentals of Acoustics. Department of Mechanical Engineering, University of Adelaide, Australia.
2. ANSI S1.1-1994, American National Standard. Acoustical Terminology.
3. Environmental Protection Department (The Government of the Hong Kong SAR). Propagation of Sound. http://www.epd.gov.hk/epd/noise_education/web/ENG_EPD_HTML/m1/intro_6.html (accessed January 29, 2013).
4. Sengpiel, E. 2010. Tontechnik Begriffe. http://www.sengpielaudio.com/TableOfSoundPressureLevels.htm (accessed January 29, 2013).
5. Minnesota Pollution Control Agency. 1999. A Guide to Noise Control in Minnesota: Acoustical Properties, Measurement, Analysis, Regulation. Revised 3/99.
6. Environmental Protection UK. 2012. Noise. http://www.environmental-protection.org.uk/noise/ (accessed January 29, 2013).
7. Occupational Safety and Health Administration. 2012. Occupational Noise Exposure. http://www.osha.gov/SLTC/noisehearingconservation/ (accessed January 29, 2013).

CHAPTER 5

End-of-Life Opportunities for Computers and Computer Parts

ABBREVIATIONS AND DEFINITIONS

CAD: Computer aided design
CDROM: Compact disc read-only memory
EOL: End of Life
EPA: U.S. Environmental Protection Agency
IT: Information technology
PC: Personal computer
PVC: Polyvinyl chloride, which is used in a wide variety of manufactured products
WEEE: Waste Electrical and Electronic Equipment Directive

5.1 INTRODUCTION

End of life (EOL) is a terminology used in product lifecycle management, mostly in the development, maintenance, and marketing fields, to signify that a product has reached the end of its useful lifetime and will no longer be marketed, sold, or sustained. In this case, the production for this product will be restricted, transferred, or ended.

Not all products have the same lifecycle curve, as this depends on many factors. Fad items, also known as short-term items, have a duration time as short as a few months. They enjoy a few months of unexpected popularity, but disappear just as quickly as they appear. On the other hand, trend or long-term products, such as a computer, have a duration that can be up to or exceed a decade.

Even with distinct lifecycle curves, the life phases of all products remain the same: Development, Introduction to the Market, Growth, Maturity, Decline, and EOL.

When a company develops a product, the primary objective is to have a long product lifecycle, in order that the company can get returns on all investments into the product's development as well as additional profit. However, with time, every product technically becomes obsolete, that is, eventually it is forced out of the market by newer products or its manufacture is stopped because the product no longer brings returns. Products live only if they survive market competition, that is, if they can satisfy and fulfill consumers' needs that other competing products do not fulfill.

In the EOL phase of a product, spare part availability is always one of the main issues for customers and manufacturers. Our suggestion in Figure 5.1 is to extend the EOL phase of the spare parts to allow the customer to repair or replace some damaged parts instead of replacing the whole product due to scarcity of parts.

Product support during the EOL phase varies from product to product. In the computing field, this has significance in the production and supportability of software and hardware products. For example, software compatibility is an important factor to consider in hardware selection because if

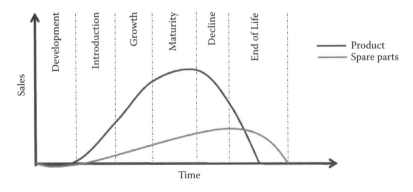

FIGURE 5.1 Lifecycle curves.

the software cannot run, then the hardware is useless. Most CAD applications always come with performance and resource compatibility requirements including disk, processor, and memory utilization. In order to run these applications, the hardware must have some specific capacities.

As mechanical engineering students, we had a project in our IT class. We were requested to adjust the parameters of a simulation tool to make it as close as possible to a real life situation. The lecturer gave us the software to install and all the information we needed for the simulation. During the installation of the software, we had some problems resulting from a damaged processor. We called the processor manufacturer and asked for a solution but were told that this type of processor was no longer manufactured and no maintenance support was available; they recommended we upgrade to a new processor. After a week, we got a new processor. After installing the software on the new processor, we were not able to boot the computer because the new processor was not compatible with the motherboard. The situation was unusual because we still had a year's manufacturer warranty but our processor was already out of the market and had no manufacturer maintenance support. In order to move on with the project, we had to buy a brand new computer.

After this project, our next "assignment" was what should we do with the old computer?

In the electronic industry, which is one of the most innovative business fields today, products based on modern technologies that enable more quality and cheaper manufacturing, less environmental pollution, etc. set trends. New products, by virtue of their high quality and lower price, force out old-fashioned products, made in classic ways, from the market. In the following, we will introduce some reflections in order to find a way toward better EOL options for computers.

In 1991, Carnegie Mellon University estimated that a minimum of approximately 150 million PCs would be buried in U.S. landfills by 2005. This number came from a study that was based on three fundamental facts: 10 years of historical U.S. sales data with a forecasted 5% future growth rate, a simple one-stage model of EOL disposition (see Figure 5.2), and some predictive assumptions on the future disposition rates of computers.

The one-stage model used in the original report (shown in Figure 5.2) is inappropriate because computers considered obsolete by their initial users often have subsequent value to others, and can stay in productive usage for a few more years.

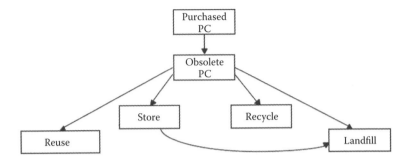

FIGURE 5.2 Flow diagram of computers based on the 1991 Carnegie Mellon University study [1].

After reconsidering the premises and considering the improvements made in product lifecycle management, especially in the EOL phase, the numbers provided by this study were revised in a new study in 1997 at the same university.

In this new study, known as Disposition and End-of-Life Options for Personal Computers [1] led by H. Scott Matthews, Francis C. McMichael, Chris T. Hendrickson, and Deanna J. Hart, sales data was updated to 1997, and it was assumed that a growth rate of 15% per year continued. They then proposed a multiple-stage model, as summarized in Figure 5.3 [1].

The arrows in Figure 5.3 define the footpaths of computers in the updated multi-stage model. A new computer is purchased by an organization, corporation, or private user. After some time, this computer becomes obsolete because it has stopped fulfilling the needs of the user. At this stage, the owner of the computer has four options. First, the computer could be

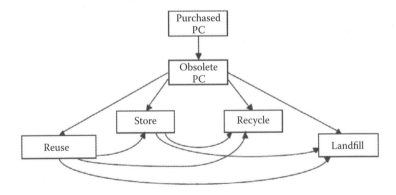

FIGURE 5.3 Flow diagram of computers in updated study [3].

reused. This means that it somehow becomes useful after becoming obsolete to the initial purchaser—possibly after being resold or reassigned to another user without extensive modification. Second, the computer could be stored by the original owner. In this case, it is serving no purpose except occupying space. Third, the computer could be recycled. In this case, the product is taken apart and individual materials or subassemblies are sold for scrap. Finally, the computer could be landfilled. In this model, the reuse and storage options are only intermediate stages in the lifecycle of a computer. Only recycling and landfilling are terminal points.

With the new updated flow diagram, we noticed a huge improvement in comparison to the old one. This new flow diagram allows a computer to be purchased, reused, stored, and finally recycled or landfilled.

5.2 BEFORE STARTING THE PROCESSES OF THE FLOW DIAGRAM

Computers are made from valuable resources and materials, including metals, plastics, and glass, all of which require energy to mine or manufacture. Donating or recycling computers conserves natural resources and avoids air and water pollution, as well as the generation of greenhouse gas emissions that are produced by manufacturing raw materials.

For example, according to the "General Information on E-Waste" published by the U.S. EPA in 2009, recycling 1 million laptops saves the energy equivalent of electricity required by more than 3647 U.S. homes in a year. Before disposing, donating, or recycling used computers or laptops, IT experts recommend that some rules be followed. Next, we examine and analyze some of these rules.

5.2.1 Delete All Personal Information from Our Electronics

For security reasons, old computers should never just be given away or disposed of without properly erasing all existing data. Simply deleting old files or reformatting the hard disk is not enough because the information is still not completely erased; only the reference to it has been deleted such that the operating system cannot find it. All the data is still there and unless it is overwritten, it can easily be recovered using file recovery software. Because the data can still be recovered from a computer with a disk drive after transfer of ownership, it is possible to unintentionally give away a lot of private information. Even if the new owner is a decent fellow, it is unlikely that anyone will want to have people reading their emails or accessing their bank account records. As mentioned previously, it is almost impossible to wipe out a hard disk manually such that it cannot be reconstructed. However, there is a way of going about this that will leave the previous computer owner feeling safe enough.

5.2.1.1 Save Your Data

First, ensure that you have an external backup of all the data in the computer.

Note: Copying all the files of the libraries as well as of "My Documents" folder or the Desktop or wherever these files are saved is not enough. Backups should also be made for all license keys or registration keys of the programs bought and installed on a computer.

A USB or a flash drive should be used to copy everything from the computer before the next step is undertaken. Once the following few steps are taken, data saved on a computer will be irrevocably lost.

5.2.1.2 Download and Run Data Destruction Software Programs (Some Are Free)

Before moving forward in this section, it is important to mention that methods of data destruction are different between private users (single person or family) and small corporations or big organizations. Corporations and organizations need to hire professional crews for erasing their data. For domestic/private users, there are a lot of data destruction software programs out there, sometimes also called data sanitization software, disk wipe software, or hard drive eraser software. These programs are application-based methods of *completely* erasing the data from a hard drive.

Darik's Boot And Nuke, usually referred to as DBAN, is the most popular free utility that will overwrite each sector on a hard drive making data unrecoverable. DBAN is freely available in a ready-to-go ISO format so all a user needs to do is download and burn it to a CD as shown in Figure 5.4 (we name our CD Darik's Boot And Nuke CD).

After deleting, restart the computer, go into the BIOS, and set it to boot from CD/DVD first. Insert the Darik's Boot and Nuke CD so the computer can boot off it to get the interface menu. The DBAN program's menu interface is also very easy to use. From this step, as represented by

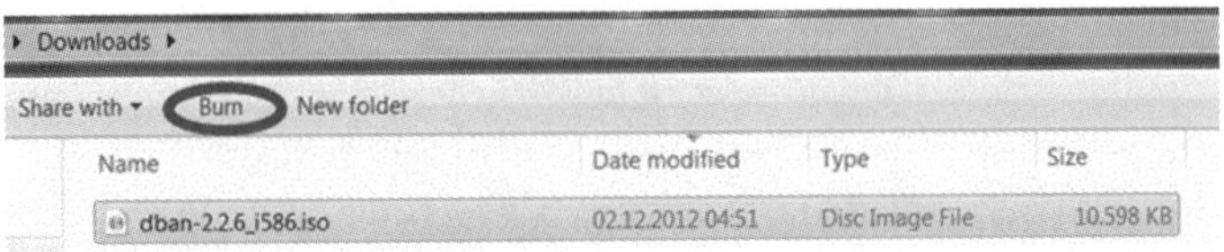

FIGURE 5.4 Downloaded ISO file.

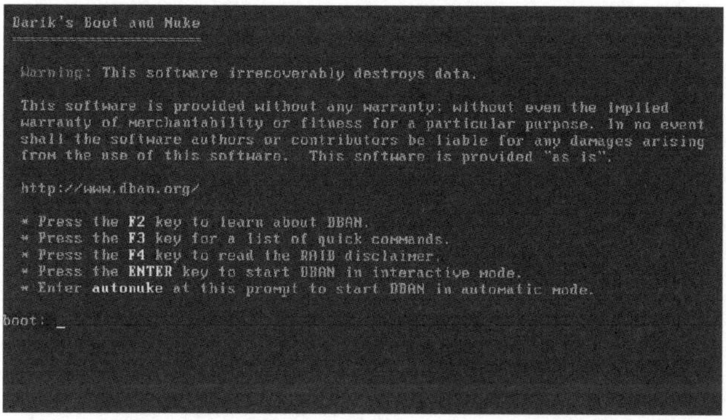

FIGURE 5.5   DBAN interface menu.

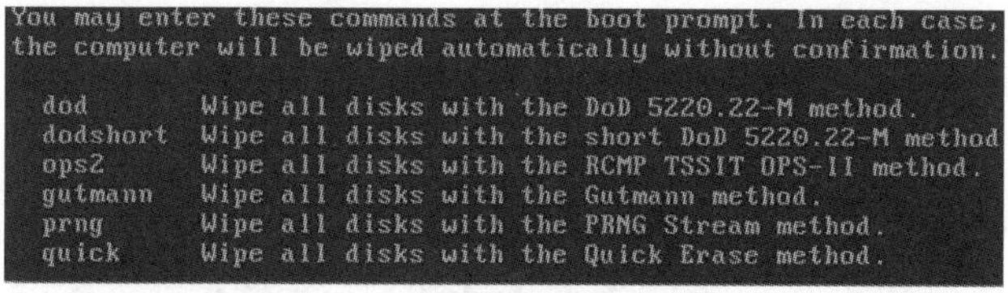

FIGURE 5.6   List of some methods used by the DBAN.

Figure 5.5, we can choose the method with which we want to wipe the disk.

By pressing the command F3 at the main menu, a list of methods is generated with which the disc can be wiped (see Figure 5.6).

5.2.1.2.1 The Autonuke Command   The easiest way to wipe a disk is to enter the command autonuke in the main interface (Figure 5.5) and press enter. This will wipe any fitted hard drives, using the default options. The progress of each hard drive will be displayed in the main lower part of the screen. The time taken and an estimate of the time remaining is displayed in the Statistics box. Figure 5.7 shows DBAN wiping two hard drives.

The process will take several hours depending on the size of the drive. When completed, DBAN will display a message similar to the one in Figure 5.8. After this, the power button on the computer can be held down in order to shut off the computer.

5.2.1.2.2 Interactive Mode   Interactive mode allows greater control. To start DBAN in interactive mode, press ENTER on the DBAN start screen (Figure 5.5). When DBAN has started, the screen in Figure 5.9 will appear.

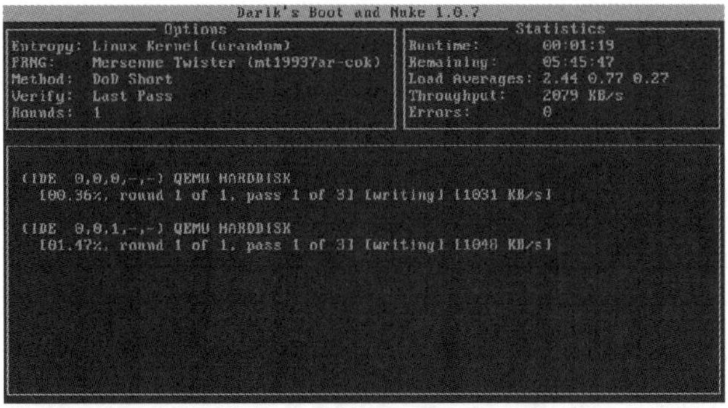

FIGURE 5.7   DBAN wipinf 2 hard drives.

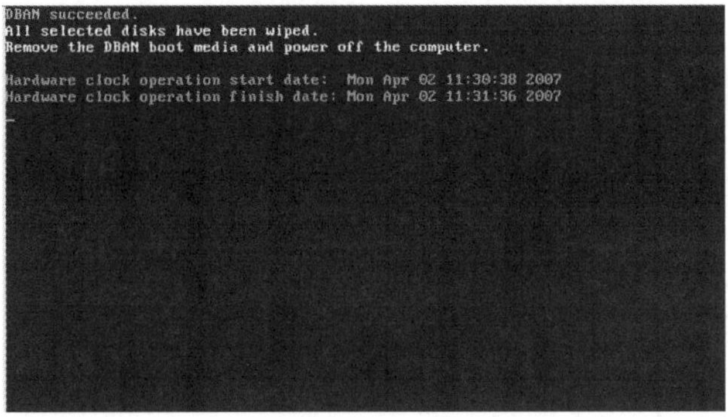

FIGURE 5.8   End of disk wiping by the DBAN.

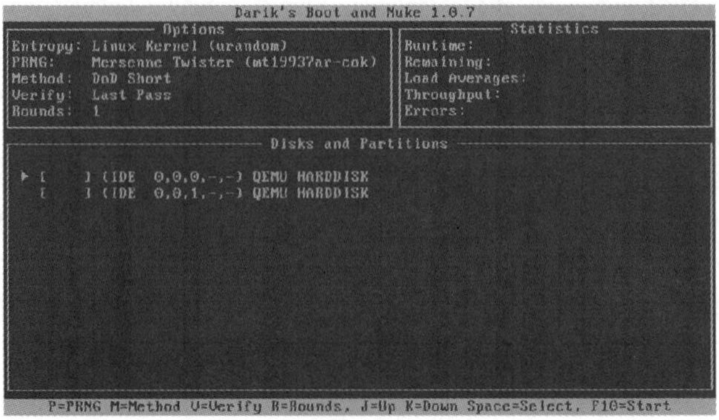

FIGURE 5.9   DBAN in interactive mode.

Use the up/down cursor and SPACE keys to select the hard drives and partitions to be wiped. If desired, select the pseudo-random number generator, wiping method, verification method, and rounds. Each option offers an informative text when it is selected. If this is appropriate for the selected option, press F10 to start wiping.

As with autonuke mode, when DBAN has finished, it will display a message, and the CD can be removed and the computer powered off.

5.2.1.3 If You Don't Trust the Data Destruction Software Programs or Your Computer Cannot Boot

The procedure described previously is well and good if the computer still works and we trust the data destruction software program. However, if this is not the case, one can remove the hard disk from the computer before disposal, donation, or recycling.

This old hard drive can be used as an external hard drive; we can also remove the powerful magnets inside the old hard drive and use them at home. The disks can be used as good signal/shaving mirrors (for camping) and the disk bearings could be used in a project, too. Alternatively, one can be creative and turn it into something funny for kids to play with or for decorative purposes in the living room like the table in Figure 5.10 [2].

FIGURE 5.10 Table in the living room. (Reprinted with permission from David Maloney (designer and builder), 2009. http://www.thenewsisbroken.com/blog/post/index/109/Latest-Project.)

5.2.2 For a Computer or Laptop, Consider Upgrading the Hardware or Software Instead of Buying a Brand New Product

In 2010, during an engineer convention in Greensboro, NC, the following topic was addressed: A person was using a 3-year-old desktop computer (running Windows XP), and this person was pondering whether to get a new computer or just upgrade the existing one. What should this person be advised to do?

Ninety percent of the students advised the person to buy a new computer because there are some aspects to a new computer that may be hard to get in an older machine. The most frequently mentioned aspects were:

- New operating system

- Faster USB port for better and faster data transfer, which also gives almost infinite flexibility and expansion

- Better wireless connection protocol with higher speed

- Some new software does not support old OS (media player, new iTunes, etc.)

- Larger and better mass storage is standard

- New warranty (which is not possible in the case of an upgrade)

The students answered the questions and defended their points without thinking of what should be the main problem here. The topic above, which was given to the students, was incomplete. Many parameters were missing. If somebody asks whether he or she should upgrade his or her computer or buy a new one, the following questions should be asked: Why is the person unhappy with the existing PC? Is it too slow? Does it have reliability problems? Does the person want to play the latest PC games, work with images from a digital camera, or edit video from a camcorder?

Very often, a few appropriately selected upgrades could help ease more advanced and suitable computing without leaving holes in our budget or negatively affecting the environment. Also, regardless of the fact that the computer companies do not like to hear this, upgrades can allow putting off purchasing a new system for months or even years.

Despite the tempting call of new ultra-fast PCs packed with the latest features and tons of storage space, many of us just do not need that much PC power. If most of our time is spent sending e-mail, surfing the Web, doing household bookkeeping, or even writing the next great novel, we really do not need the flying speed fancy bells and whistles of the latest models.

5.2.2.1 How Old Is Too Old?

Before looking at the types of PC upgrades that can be taken into consideration, it is important to talk about which computers are worth upgrading. The best gauge is the age of your PC. If the computer is between 2 and 4 years old, it is a good candidate for upgrades. If the PC is much older, there is really no point in trying to upgrade it. PCs that are 4, 5, or more years old are simply unable to take advantage of the newest components such as hard drives or graphics cards.

In some cases, specific upgrades—such as processors—are not available for older PCs; in others, they will work, but at slower speeds. For example, while we can install one of the newest mega-space hard drives in an old PC, it will not work at the maximum speed. Some older PCs also require special upgrade components—such as memory chips—that are difficult to find or so expensive that upgrades just are not economical.

5.2.2.2 When to Buy a New PC

Apart from the age of the computer, there is no firmly established rule for when upgrades are not worth it. However, if we decide to upgrade most of our PCs components with higher-end options, the price can quickly come close to the cost of a brand new PC. In that case, it may be better to opt for a new computer, which will give us a system where everything is designed to work together using the latest technology.

In addition, some applications require superfast systems. This is particularly true for editing video from camcorders or playing the latest eye-popping computer games. These require very fast processors, a lot of RAM, and big high-performance hard drives. Admittedly, we can get by with an upgraded system, but we will have to live with the compromises. For example, we can edit video on an older, upgraded system, but we will spend time twiddling our thumbs while the system catches up with us. We will not be able to use some of the more advanced video effects that would be a piece of cake for a brand new system.

5.2.2.3 Moving Data to a New PC

If you end up opting to buy a new PC, you have probably wondered how to move your data from the old PC to the new one. It is a major concern, but there are possibilities. If your new PC comes with Microsoft Windows

FIGURE 5.11   Windows 7 Easy Transfer.

7 already installed, as is likely the case, Windows 7 includes a "Files and Settings Transfer Wizard" that will help copy essential data from your old PC. (On Windows 7, click Start, All Programs, Accessories, System Tools, and then choose Windows Easy Transfer. Then, you will see a window like the one shown in Figure 5.11.)

If you do not have a home network, you will need an Easy Transfer cable (approximately $40) to connect your old and new computers (Figure 5.12).

However, one thing the Window Easy Transfer Wizard does not do is transfer applications from the old to the new computer. You will need to

FIGURE 5.12   Easy Transfer Cable.

reinstall them. To do the whole job, you can use a program like PC mover professional ($60), which transfers all applications in addition to files and settings.

5.2.2.4 Choosing the Upgrade Path

Upgrading a computer is not rocket science. If you are equipped with a screwdriver, you can do it. However, it does require some careful, methodical work. If you are all thumbs, or the impatient type, you can still forge ahead on the upgrade path, but it is not a bad idea to get a computer-savvy friend to help. Below, we have listed the most popular PC upgrades.

Warning: Do not expect to be able to upgrade your PC's microprocessor. Two to three years ago, processor upgrades from companies such as Kingston and Evergreen Technologies were readily available and popular. However, they are usually not an option anymore because today's motherboards are designed for specific processors running at specific speeds. Moreover, even if we could upgrade to a faster CPU, we would find that it would not make a huge difference in overall system speed. The processor is just one of the many components in a PC that must work together for maximum performance.

5.2.2.4.1 Memory   Memory upgrade is one upgrade that can really affect the overall performance of your computer. Increasing the computer's memory can help increase the speed and help with load times of many programs. In addition to being an effective upgrade, it is an affordable, often easy, and fast form of upgrade.

If your computer could use a small performance boost and has less than 1 GB of memory, this can be an effective upgrade and you may want to do this instead of purchasing a new computer.

Upgrading your PCs memory remains one of the most effective—and most economical—ways to bump up the PCs performance. It is also one of the most popular upgrades. Today's operating systems and applications run faster with large amounts of RAM, but if your PC is a year or two old, chances are that it came with just 2 GB or 4 GB of memory. Equipping such a PC with 8 GB of RAM (approximately $100) will make a noticeable difference in overall speed, especially with today's memory-hungry applications such as graphics-intensive games. In addition, a PC with more memory is less likely to lock up or behave strangely.

Depending on the design of your PC, you may be able to add to the memory you have, or you may need to discard what you have and start from scratch.

5.2.2.4.2 Hard Drive   In the past, upgrading the hard drive was a popular option. However, today hard drives are becoming so large that most users are not running out of disk space. However, if your computer is running out of hard drive space, you are happy with the performance of the computer, and considering the options between upgrading and purchasing a new computer, adding a new hard drive to the computer is a cheap and often simple upgrade option. After RAM, hard drives are the next most popular PC upgrade. Drives keep getting bigger and prices keep falling.

5.2.2.4.3 DVD/CD-RW Drive   Adding a new disc drive such as a faster CD-R/CD-RW (CD burner) or a DVD drive can be a great upgrade to a computer that either has a slower drive or does not have a CD-ROM/ CD-R/CD-RW/DVD.

If you burn or create many of your own discs, a new burner can greatly decrease the time it takes to create each disc. In addition, a faster drive can also help with the installation process or load time of any CD or DVD programs. Apart from these two examples, this upgrade is more of a new feature than a performance upgrade.

This upgrade adds real utility to your computer. You can share digital camera photos with friends and family, secure backups of your important data, or create your own music mixes. And that is just the beginning.

5.2.2.4.4 Video Card   If you play games on your computer, an upgraded video could be one of the best upgrades for your computer. With the new improvements and complexities of graphics in games, many older computers or computers with inefficient video cards are slow.

Adding a new video card is often an easy upgrade and is a lot cheaper than purchasing a new computer.

5.2.2.4.5 Graphics Card   If you work with photos from a digital camera or play PC games, a new graphics card cannot only make images pop up on your screen faster, but can produce sharper, higher-resolution images that are easier on your eyes. Thus, a new graphics card is an easy upgrade that pays off.

5.2.2.4.6 Sound System   If you are into PC music, a new sound card and the latest-technology speakers make a difference that you can hear. Numerous choices are available, from inexpensive to wallet-emptying, but spending $75 to $100 for a new sound card and $50 to $100 for speakers can result in sound quality that rivals the stereo system in your living room.

5.2.2.4.7 Monitor   Not every upgrade requires opening the PC case. Since we spend all our PC time looking at our monitor, investing in a newer, bigger monitor can be better than getting a new PC. Our eyes will thank us for this.

Although a new display is not going to improve the performance of a computer, adding an LCD or flat-panel display or increasing the size of the display can make using the computer a much more enjoyable experience.

If your display is 17 in. or smaller, you are using a monitor and not a flat-panel display, and you have the room to do it, this upgrade is easy and a lot less expensive than purchasing a new computer. In addition, if somewhere down the road you do purchase a new computer, you will not need a new display.

One of the best things about a monitor upgrade is that you can continue to use it. When you finally do buy that new PC, just plug it in.

5.2.2.4.8 Broadband   Surfing on the Internet is one of the most popular things people do while on the computer and many users often do not think about their Internet connection when considering upgrading their computer. If you do not have a broadband Internet connection available in your area and you spend a lot of time on the Internet, subscribing to a broadband service would be a lot more beneficial than doing any other upgrade or purchasing a new computer.

Users who already have broadband may also consider faster broadband solutions. For example, if you have DSL you may want to consider changing to cable or another faster solution.

5.2.2.4.9 Other Upgrades   Apart from the previous recommendations, if you are looking at doing any other upgrades such as motherboard, modem, network card, etc. today it is almost always recommended that you save your money and purchase a new computer now or at some time down the road.

5.2.3 Remove Any Batteries from Your Electronics, They May Need to Be Recycled Separately

Which kind of batteries are in a computer system?

5.2.3.1 CMOS Battery

CMOS (complementary metal-oxide-semiconductor) is the term usually used to describe the small amount of memory on a computer motherboard that stores the BIOS settings.

The CMOS is usually powered by a CR2032 cell battery as represented by Figure 5.13. Most CMOS batteries will last the lifetime of a motherboard but will sometimes need to be replaced. Incorrect or slow system date and time and loss of BIOS settings are major signs of a dead or dying CMOS battery.

In many cases, the battery is soldered directly onto the motherboard, but the battery is usually in some sort of a holder so it is easy to replace it. Computers are not the only gadgets that have small batteries—camcorders and digital cameras often have them, too. Just about any gadget that keeps track of time has a battery.

In a computer (as well as other gadgets), the battery powers a chip called the **Real Time Clock (RTC)** chip. The RTC is essentially a quartz watch

FIGURE 5.13   CR2032 CMOS battery.

that runs all the time, whether or not the computer has power. The battery powers this clock. When the computer boots up, part of the process is to query the RTC to get the correct time and date. A little quartz clock like this might run for 5 to 7 years off a small battery.

This does not explain why your computer would not boot, however. We would expect the computer to boot fine but have an incorrect time and date. The reason your computer would not boot is that the RTC chip also contains 64 (or more) bytes of RAM. The clock uses 10 bytes of this space, leaving 54 bytes for other purposes. The BIOS stores all sorts of information in the CMOS RAM area, like the number of floppy and hard disk drives, the hard disk drive type, etc. If the CMOS RAM loses power, the computer will not know anything about the hard disk configuration of your machine, and therefore it cannot boot.

Many computers that are more modern are not so dependent on the CMOS RAM. They store the settings in **non-volatile RAM** that works without any power at all. If the battery dies, the clock fails but the computer can still boot using the information in the non-volatile RAM area.

5.2.3.2 Laptop Battery

Laptops are powered by several different types of batteries. These batteries provide direct-current power to your laptop. Figure 5.14 represents a high quality Dell Inspiron 1520 battery 4400 mAh.

According to Laptop Travel, a U.S.-based company that offers solutions for international laptop connectivity, there are three different types of laptop battery chemistry. Nickel cadmium was the first chemistry, followed by nickel metal hydride and lithium ion battery types. Notebook batteries

FIGURE 5.14   Laptop battery.

are also sometimes referred to as "smart" or "dumb." These terms are explained next.

5.2.3.2.1 Nickel Cadmium (NiCd)   NiCd batteries were the first rechargeable batteries for notebook computers and featured low cost, versatility, and high output current capability. NiCd batteries can be charged rapidly and used in a wide range of products. However, they are now being designed out of notebooks in favor of the newer and higher power NiMH and LiON batteries.

5.2.3.2.2 Nickel Metal Hydride (NiMH)   NiMH batteries represented a significant improvement over NiCd batteries due to improvements in cost, safety, reliability, and capacity. Currently the most widely used notebook battery, the NiMH battery has only one drawback. The "memory effect" of NiMH batteries requires that they be fully discharged prior to recharging for maximum charge effectiveness.

5.2.3.2.3 Lithium Ion (LiON)   LiON batteries are now the most popular notebook battery because of improvements over NiMH in the area of memory effect. In addition, for batteries of comparable capacity, LiON batteries are generally somewhat lighter in weight than NiMH batteries. Because the LiON chemistry is the newest in the market and offers these benefits, the market price for LiON batteries is often significantly higher than for NiMH batteries with the same actual capacity.

5.2.3.2.4 Smart/Dumb Batteries   Some batteries are classified as "smart" or "dumb." Smart batteries have internal microprocessor circuits that help manage battery energy, report the state of the charge, predict running time, and track battery usage. Dumb batteries are counterparts of smart batteries but do not have these smart battery features [3].

When disposing, laptop batteries must be removed as well as the CMOS battery on the motherboard, as many batteries are considered hazardous waste. Most CMOS batteries today are lithium, which can be recycled for clean scrap metal as well as lithium carbonate, which, according to battery recyclers, can be used in making new batteries. Laptop batteries may be nickel metal hydride, nickel-cadmium, or lithium-ion, all of which can be recycled for metal oxide.

5.3 THE FLOW DIAGRAM

5.3.1 Computer Donation/Reuse

Just because we no longer have use for an old desktop or notebook PC does not mean somebody else will not find it usable. When a computer stops fulfilling the needs of its users, it can be donated to a nonprofit organization or directly to someone who needs it the way it is. For example, technical sales always need powerful notebooks for their demonstrations in front of the customer. After 2 years, they must change their laptops in order to be in sync with the applications development. Their old laptops can then be reused by office assistants, who will need the laptop more for e-mail, PowerPoint presentation, and Excel table management.

5.3.1.1 How to Reuse

It is always better to donate a good running computer with an operating system installed.

5.3.1.2 Where to Donate

We may be tempted to donate equipment directly to a favorite local school or charity. However, keep in mind that most organizations have very specific technology needs. A donated computer might not be a good gift. Refurbishers or nonprofit organizations are better equipped to repair and upgrade older computers. They will ensure that the equipment works well and runs legal software copies and that any e-waste is disposed of properly. They will pass on ready-to-use equipment to those who need it, often at little or no cost to the recipient.

5.3.2 Computer Storage

Perhaps you have an old computer you no longer want but have not decided whether to donate or recycle it. Maybe you want to salvage some components for yourself or for resale and then donate what is left to a charity or a friend. Whatever the case, a computer—in whole or in part—will be most useful to its next owner if it is properly packaged and stored.

According to the document "Solid Waste and Emergency Response (5306W) EPA530-N-00-007" [4] from the EPA, it is estimated that 20 million PCs became obsolete in 1998. Most of these are in storage—meaning that they are in closets or storerooms in homes, offices, and businesses. The sooner computers get out of storage and into the recycling channel the more likely it is that they can be used again for their original purpose. However, even the oldest and most damaged computers can be recycled because many contain parts that can be refurbished and reused such as metals, plastics, glass, and other materials that require energy to be obtained and manufactured. When we store old computers, we are also storing resources that could have been used elsewhere, therefore generating additional unwanted environmental damage.

5.3.2.1 How to Store

In order to store a computer in the proper way, here are some steps that should be followed: Gather all of the components to one section of a room, unplug cables from each port on the back of the computer, and wipe down any debris on the outer casing. Access the inside of the computer from the side panel to vacuum dust. Decide which components (video card, sound card, wireless card, or motherboard) you want to remove from any internal slots. Remove selected items carefully. Wrap these in a bubble wrap. If you have static-resistant bags, put these accessories inside (Figures 5.15 and 5.16) to prevent static charge from rendering them useless. Close up the computer after all necessary components are removed.

Begin wrapping the internal components in bubble wrap as extra precaution. Bubble-wrap the keyboard, mouse, speakers, and all accessories.

Continue wrapping the computer and monitor (Figure 5.17). Set up a box for each bulky item such as the computer and monitor. Place the hardware into the box and surround it with newspaper. Seal the box and write

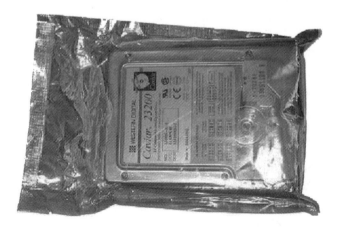

FIGURE 5.15   Desktop computer hard drive in an antistatic bag.

FIGURE 5.16   Laptop hard drive in an antistatic bag.

FIGURE 5.17   Wrapping a monitor and keyboard.

on the outside which components it contains. Put internal hardware into one box and smaller accessories into another. Place filled boxes in a cool storage area.

5.3.2.2 Where to Store
If you have a garage, basement, or hobby room in your home, these would be good places to store the boxes containing your computer parts. You can also store your computer in a room that is not exposed to the sunlight and out of reach of children.

5.3.3 Computer Recycled

Electronic equipment contains materials that can be hazardous to human health and the environment if they are not properly managed. Computer monitors contain an average of 4 lb of lead and require special handling at the end of their lives. Computers and related electronic equipment can also contain substances such as metals, glass, plastics, chromium, cadmium, mercury, beryllium, nickel, zinc, brominated flame-retardants, and certain chemical compounds that are highly recoverable, recyclable, and reusable. By recycling our old equipment in a responsible manner, we can keep electronic products out of our landfills and help maximize the use of natural resources.

According to the EPA document "Electronics Waste Management in the United States Through 2009," [5] 25% of electronics were collected for recycling in 2009; and in Table 5.1 we can see that computers have been collected at the highest rate, which is 38%.

5.3.3.1 How to Recycle

Many computers are built to be easily taken apart into their component parts for easy recycling. Some devices may require more energy to recycle, but it is still better than tossing them into a landfill. The European Union Waste Electrical and Electronic Equipment (WEEE) Directive (Recast) sets the criteria for correct disposal of electronic waste, and while householders can refrain from compliance schemes to dispose of WEEE, companies in a business-to-business environment have a legal responsibility to dispose of obsolete electronic equipment according to the directive. In Ireland, for example, this is covered by the Irish WEEE Regulations of 2005. IT equipment, when handled properly, is almost 100% recyclable and service providers should be able to provide a full downstream report of the final destination of all of the different materials contained in the recycled equipment.

TABLE 5.1 Rate at Which Used Electronics Are Collected for Recycling Relative to the Total Weight of Each Product Ready for End-of-Life Management, 2006 to 2009

Calendar Year	Computers	Computer Displays	Hard-Copy Devices	Keyboards and Mice	TVs	Mobile devices	Total
2006	33%	21%	37%	7%	16%	6%	22%
2007	36%	24%	38%	7%	17%	7%	24%
2008	38%	26%	35%	7%	16%	11%	24%
2009	38%	29%	34%	8%	17%	8%	25%

5.3.3.2 Where to Recycle

Creating and managing your own computer-recycling program can be a costly exercise for a private person or a small business. It is better to use a licensed recycler who can do it at a reasonable price and manage environmental responsibilities at the same time.

In doing so, make sure that the company you choose is properly licensed to recycle computers. Request a certificate of Recycling/Destruction.

Electronics recycling is a new industry, and it is far from centralized at this point. Many people end up throwing their old electronics in the trash simply out of frustration. It can take a good deal of research to figure out how to properly recycle this stuff. Going to the manufacturer's Web site or to the store where the device was purchased is often a good bet. Many electronics manufacturers and retailers have instituted collection programs that make it easy to recycle old gadgets. If these programs are not available in your city, then you might want to call your city council and ask.

5.3.4 Computer Landfilled

According to Greenpeace [6], the U.S. EPA states that more than 4.6 million tons of e-waste ended up in U.S. landfills in 2000. Chemicals like lead, PVC, retardants, chromium, mercury, beryllium, and cadmium are all found in a single desktop and monitor system. When placed in a landfill, these toxins can leak into the ground over time or are released into the atmosphere and pose severe health risks such as damage to the brain, liver, kidneys, lungs, and reproduction system. Exposure to these toxins can also harm fetal development. In many European countries, regulations have been introduced to prevent electronic waste from being dumped in landfills due to its content of toxic heavy metals. However, the practice continues in many countries. In Hong Kong, for example, according to the same Greenpeace article, it is estimated that 10 to 20% of discarded computers end up in landfills.

E-waste that does not remain in a landfill and is incinerated instead releases heavy metals such as lead, cadmium, and mercury into the air. Mercury released into the atmosphere can accumulate in the food chain, particularly in fish—the major route of exposure for the public. If the products contain PVC plastic, highly toxic dioxins and furans are also released. Brominated flame-retardants generate brominated dioxins and furans when e-waste is burned.

As if the above is not bad enough, there is a more alarming concern from the rise of shipping e-waste to developing countries for processing.

In poor developing countries where e-waste is shipped, landfill workers are often children, women, and the elderly, who work to extract the toxic metals with simple tools like hammers and pliers, wearing no protective gear of any sort, and then discard the parts, recycled or not, in open spaces completely exposed to the elements. Such an approach endangers the lives of people and contaminates the environment.

Although reuse and recycle are positive approaches to protecting the earth from unwanted trash, the shipping of e-waste from developed nations to developing countries needs better regulation; otherwise, the pollution simply moves to another country.

5.3.4.1 How to Dispose

The metals in computers and monitors, including copper, gold, and aluminum, are recyclable and could be used to make new products. Instead, millions of pounds of these metals are in landfills unused. Several tons of cadmium from computers is dumped into landfills every year. These metals still have useful life and may be recycled into a variety of new products that contain the same metals. In addition, glass from computer monitors can be reclaimed and made into new glass products. Even the computers and monitors themselves may contain components including hard drives and CD-ROM drives that are still working and can be used to assemble computers for charity organizations, schools, and libraries. It is important to keep computers and monitors out of landfills and to dispose of them where they will be recycled and toxic materials will be appropriately handled.

5.3.4.2 Where to Dispose

With millions of desktop and laptop computers sold in the world every year, these machines have the potential to take up hundreds of millions of cubic feet of landfill space at the end of their useful life. If proper attention and care are not given to this issue, more money will be spent on medical bills, there will be less available space as playgrounds for children, and more dumping grounds will arise. In some countries, dumping computers in a landfill is against the law. People and businesses with computers and printer equipment that is past the end of its useful life should check with their local government to find out what the requirements are for electronics disposal to reduce harm to the environment, to avoid paying fines for improper disposal, or risk legal action from people affected by the toxic substances leeched from improperly disposed computers.

5.3.4.3 Agbogbloshie Computer Graveyard in Ghana

Agbogbloshie is a suburb of Accra, Ghana known as a destination for legal and illegal exportation and hazardous dumping of computer waste from Western nations.

When we arrived in Agbogbloshie, the first intriguing thing we saw was a container that was about to be opened. On the container was the following label: charity donation—second-hand computers. However, in reality as much as 80% of the computers in the container were damaged or obsolete.

Many young people from the rural areas of Ghana work here unprotected, mostly poorly paid boys between 11 and 18 years old as shown in Figure 5.18. They dismantle and break computers into pieces, and melt down circuit boards, mainly to extract the copper, aluminum, and iron, which they collect and sell. The business has created massive pollution from leaded glass and other toxic materials. In 2008, Greenpeace sampled the burned soil at Agbogbloshie and found high levels of lead, cadmium, antimony, PCBs, and chlorinated dioxins. Such pollution could be

FIGURE 5.18 Computer graveyard in Agbogbloshie, Ghana.

mitigated by moves to recycle and properly dispose of so-called electronic wastes, which are gaining ground in the West. A European Union law requires manufacturers to recycle junk electronics free of charge, although policies in the United States are fragmented in different areas.

5.4 CONCLUSION

A more interesting statistic than the number of computers recycled or landfilled is the number that is available for product take-back. Product take-back is an emerging international paradigm that requires firms to organize methods to reclaim their products at the end of their useful life. According to the WEEE directive [7] set by the European parliament and council, in cases where a final end-user purchased products directly from a manufacturer or distributors, they should arrange for the recovery and recycling of the product at no additional cost. In such cases, once the WEEE is collected, the manufacturer or distributor will work with their designated suppliers to ensure the latest available technologies and methods are used to safely and efficiently recycle the products, and that as much as possible reusable materials are recovered. Final electronic waste will be disposed in the most environmentally friendly manner.

The 27 European countries making up the European Union have some form of take-back legislation based on WEEE enacted already, and firms are becoming increasingly aware of the possibility that they will be required to comply. In this model, the number available for take-back is defined as those that either have been landfilled or are currently in storage. A 1997 study carried out by Carnegie Mellon University, Matthews and his colleagues predicted that in 2005, despite increased recycling efforts, 80 million computers would still not be available for take-back [1].

To more completely understand the environmental impacts of PC systems, they should be broken down into their major subcomponents: motherboards, cases, drives, monitors, and user input devices. This breakdown is basically unchanged since 1991, other than the fact that CD-ROM drives are now a standard part of all systems. This is the definition used in subsequent analysis.

It has probably become apparent to the reader that storing obsolete computers is inadvisable because their value will only decrease over time, until they are worth only the sum of their raw materials. Unfortunately, because the residual value is in competition with wholesale prices, the retail value of used IT equipment is substantially lower than the original cost. The residual value of materials in old electronic equipment soon after production is only 1 to 5% of the original cost of the equipment [1]. Storing an old computer instead of quickly reusing or recycling it is not a profit-maximizing or cost-effective solution. Holding on to an obsolete computer without getting some benefit out of it is akin to holding on to a stock of a company that is known to be going bankrupt—in either case, we end up with nothing and have full knowledge of this fact ahead of time.

Further complicating the issue of storage is that many computers that are beyond anyone's definition of "useful life" can only be disassembled for raw materials. Often, choosing this option requires paying a fee to a computer recycler just to comply with local regulations banning disposal. Reuse and recycling need to be more forcefully promoted as options to maximize lifecycle values of computers.

REFERENCES

1. Matthews, H.S., Hendrickson, C., McMichael, F., and Hart, D. 1997. Disposition and End-of-Life Options for Personal Computers. Green Design Initiative Technical Report #97-10, Carnegie Mellon University. http://www.ce.cmu.edu/greendesign/comprec/NEWREPORT.PDF (accessed January 22, 2013).
2. Maloney, D. 2009. http://www.thenewsisbroken.com/blog/post/index/109/Latest-Project (accessed December 3, 2012).
3. Yonga, J. Laptop battery types. http://www.laptoptravel.com/Article.aspx?ID=214 (accessed December 3, 2012).
4. U.S. EPA. 2000. Solid Waste and Emergency Response (5306W) EPA530-N-00-007. http://www.epa.gov/wastes/conserve/smm/wastewise/pubs/wwupda14.txt
5. U.S. EPA. 2011. Electronics Waste Management in the United States through 2009. http://www.epa.gov/epawaste/conserve/materials/ecycling/docs/summarybaselinereport2011.pdf
6. Greenpeace. 2009. Where does e-waste end up? http://www.greenpeace.org/international/en/campaigns/toxics/electronics/the-e-waste-problem/where-does-e-waste-end-up/#a2
7. Directive 2002/96/EC of the European Parliament and of the council of 27 January 2003 on waste electrical and electronic equipment (WEEE) and Directive 2003/108/EC of the European Parliament and of the council of December 8, 2003.

CHAPTER 6

Green Cloud Computing

ABBREVIATIONS

API: Application programming interface
CAPEX: Capital expenditure
CAPTCHA: Completely automated public Turing test to tell computers and humans apart
IaaS: Infrastructure as a service
IT: Information technology
MMPOG: Massively multiplayer online games
NIST: National Institute of Standards and Technology
OPEX: Operational expenditure
PaaS: Platform as a service
QoS: Quality of service
RAID: Redundant array of independent disks
ROI: Return on investment
SaaS: Software as a service
UPS: Uninterruptible power supply
WPA: WiFi protected access

6.1 BEFORE CLOUD COMPUTING

The computing world has not always been so practical, powerful, and handy as it may be the case today. First generation (1945–1958) and second generation (1958–1964) computers were mostly huge systems. Examples of first- and second-generation computers are displayed in Figure 6.1 and Figure 6.2. They were owned by corporations, universities, and other big institutions. They could not be used by common

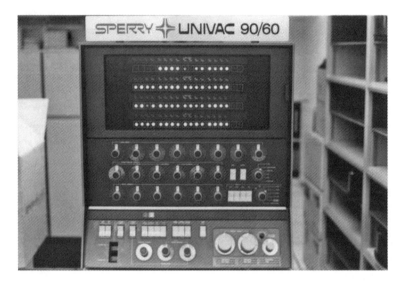

FIGURE 6.1 The console of a UNIVAC 90/60 mainframe computer (public domain).

persons because experienced skills were needed. However, there was no direct human-machine interaction. Users prepared tasks for the computer using cardpunches. These cards contained instructions for the machine and were processed in a batch mode. When the job was done, users had to collect the output. It could take days, even weeks between a job assignment and the retrieval of results.

Computer use increased with third generation (1964–1974) computers due to the invention of integrated circuits, which allowed for the development of much smaller computers (minicomputers, Figure 6.2) and as such brought computing to more people.

The creation of microprocessors gave life to fourth generation (1974–today) computers. Unlike third generation computers, which were basically a scaled down version of mainframe computers, microprocessor-based computers differed in the targeted market. The first personal computer was built using microprocessor technology.

Mainframe computers are indeed powerful but they do not fit in a bedroom or on a work desk. Additionally, not every corporation wishes to invest financial resources into purchasing a mainframe. The alternative solution is the joint use of several similar resources in order to achieve one goal.

Look at distributed compiling, for example. If you are a software developer, you will surely agree that it can be very frustrating when you have to

FIGURE 6.2 K-202 minicomputer built in 1972 in Poland by Jacek Karpiński (public domain).

wait for hours in front of your desk until the software project you are working on is successfully built. To reduce waiting time, it can be quite useful to rely on distributed building software. The idea behind this is to use both local and network computers in order to speed up the building process.

Besides distributed building, we can also involve distributed computing (or grid computing), which involves the distribution of long-running computing processes (like complex financial calculations) to several computers (grid nodes) located in the network. This speeds up the overall calculation time (compared to the calculation time on a single machine) and allows the user to get accurate results in less time.

In both previous examples, the need for availability of a network computer is imperative. In other cases, specified resources are needed only at a given point of the day, month, or even year.

Let us say we intend to launch a web shop specialized in Christmas gifts. We can assume that the shop will likely be visited between November and January. Do we need to carry IT costs (purchasing, maintenance, etc.) in order to guarantee the availability of our shop all year round, knowing that the services will not be used to their full potential?

This example is one of the multiple usage scenarios that make it worthwhile to take a deeper look into Cloud Computing.

6.2 UNDERSTANDING CLOUD COMPUTING

According to the National Institute of Standards and Technology (NIST), "Cloud Computing is a model for enabling convenient, on-demand network access to a shared pool of configurable computing resources (e.g., networks, servers, storage, applications and services) that can be rapidly provisioned and released with minimal management effort or service provider interaction." Let us illustrate this with an example. Figure 6.3 represents common email access via a laptop or a desktop computer.

In this case, several users need to access their email stored at the server "myemail.com." To achieve this, they can connect to the mail provider using any device with a network (mostly Internet) connection. From the user's perspective, it does not matter where and how the data (emails) are actually stored. The emails just need to be available on demand within a reasonable response time. The cloud symbol denotes the demarcation point between the user responsibility and the provider responsibility.

From the provider's perspective, things look a bit different. In order to offer an attractive, reliable, and available mail service, traditional providers relied on standard IT infrastructure. However, today, depending on the services provided, it can be beneficial to explore the Cloud Computing alternative.

With the deployment of IT infrastructure and IT services via a network, it is possible for organizations to purchase only what they need when they need it (on demand). With this model, it is possible to keep IT costs in view and handle changes in demand in a comfortable manner.

Figure 6.4 and Figure 6.5 illustrate the transition from traditional computing (IT infrastructure is owned, installed, and managed by individual companies) to Cloud Computing (IT infrastructure is owned by a cloud provider and shared by different independent parties/customers).

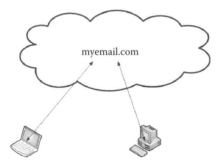

FIGURE 6.3   A simple email access.

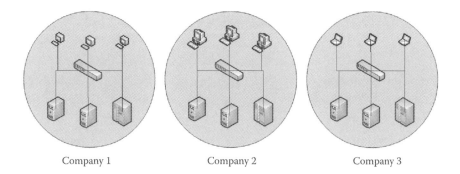

FIGURE 6.4 Traditional computing—each company owns its infrastructure.

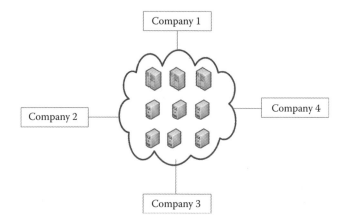

FIGURE 6.5 Cloud computing—resources are shared by different parties over networks.

Several major aspects of cloud systems are illustrated in Figure 6.6.

6.2.1 Essential Characteristics

Let us go through the essential characteristics of Cloud Computing according to the NIST definition [1]:

6.2.1.1 On-Demand Self-Service

"A consumer can unilaterally provision computing capabilities, such as server time and network storage, as needed automatically without requiring human interaction with each service provider."

Since companies are using the cloud for vital tasks, it is essential that the purchased services and resources are always available. A consumer must be able to access them without interaction with the cloud provider. In order to

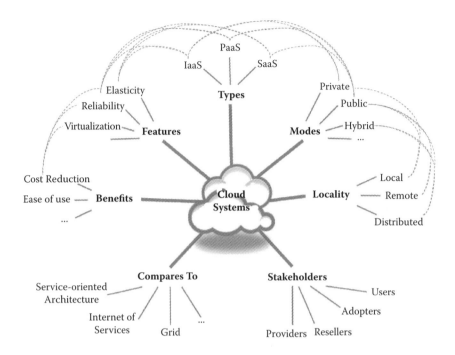

FIGURE 6.6 Non-exhaustive view on the main aspects forming a cloud system [3].

request a new server, for example, it is no longer necessary to go through a cumbersome internal administrative process. You can just browse through your provider's self-service portal and request the resource of your choice via a web form. The self-service portal allocates and configures the required resource according to your request and you can access it afterward.

This simple scenario looks great, but there are some security issues that need to be addressed. For example, how are you sure that resources that have been allocated to you are only accessible by you? From the provider's view, it must be possible to control and monitor the use of cloud resources.

Cloud customers can also rely on so-called auto-scaling services, which are given by cloud providers and can be used to modify the amount of used resources at run time. This allows customers to influence their total resources costs and providers to serve more customers using the same infrastructure.

6.2.1.2 Broad Network Access

"Capabilities are available over the network and accessed through standard mechanisms that promote use by heterogeneous thin or thick client platforms (e.g., mobile phones, tablets, laptops, and workstations)."

In fact, Cloud Computing is based on the Internet. Therefore, it is accessible from any device that has a network interface.

From the provider's point of view, the challenge here would be to make sure that a defined security level always applies no matter which device is connected to the cloud.

6.2.1.3 Resource Pooling

"The provider's computing resources are pooled to serve multiple consumers using a multi-tenant model, with different physical and virtual resources dynamically assigned and reassigned according to consumer demand. There is a sense of location independence in that the customer generally has no control or knowledge over the exact location of the provided resources but may be able to specify location at a higher level of abstraction (e.g., country, state, or datacenter). Examples of resources include storage, processing, memory, and network bandwidth."

Figure 6.7 illustrates this aspect where clients can request resources (e.g., a new server) from the provider. Which server is allocated to each client's request does not matter. For startup IT organizations, this is a great benefit because there is no need to invest in their own data center (which may not always be under full load every time). Resource pooling allows cloud providers to maximize the load on the shared resources.

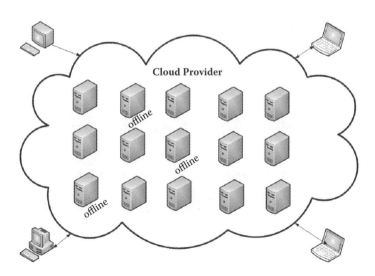

FIGURE 6.7   Resource pooling.

6.2.1.4 Rapid Elasticity

"Capabilities can be elastically provisioned and released, in some cases automatically, to scale rapidly outward and inward commensurate with demand. To the consumer, the capabilities available for provisioning often appear to be unlimited and can be appropriated in any quantity at any time."

The provider can, for example, extend the resource pool in order to fulfill growing demands easily or reduce the pool (e.g., for maintenance operations) without this being noticed by the customer.

6.2.1.5 Measured Service

"Cloud systems automatically control and optimize resource use by leveraging a metering capability at some level of abstraction appropriate to the type of service (e.g., storage, processing, bandwidth, and active user accounts). Resource usage can be monitored, controlled, and reported, providing transparency for both the provider and consumer of the utilized service."

6.2.2 Specific Capabilities of Clouds

The characteristics previously described can also be classified into economic, non-functional, and technological capabilities, which are addressed by cloud systems (see Figure 6.8).

6.2.2.1 Economic Aspects

One of the major reasons why cloud systems are (or should be) introduced in business environment is economic considerations. Infrastructure outsourcing and the possibility to automate resource management lead to a reduction of cost and effort as well. Unfortunately, infrastructure

Economic	**Non-functional**	**Technological**
• Cost reduction • Pay per use • Improved time to market • Return on Investment • CAPEX -> OPEX • Going green	• Elasticity • Reliability • Agility and Adaptability • Availability of services and data	• Virtualization • Multi-tenancy • Security, Privacy, and Compliance • Data Management • API/Programming enhancements • Metering • Tool

FIGURE 6.8 Specific categories of cloud systems.

outsourcing leads to a loss of control. IT decision makers must carefully balance the loss of control with the cost and effort reduction during the decision process.

In order to allow for economic considerations, cloud systems should help in realizing the following aspects [3]:

- Cost reduction is one of the first concerns when building up a cloud system that can adapt to changing consumer behavior and reduce cost for infrastructure maintenance and acquisition. Scalability and pay per use are essential aspects of this issue. Notably, setting up a cloud system typically entails additional costs—be it from adapting the business logic to the cloud host specific interfaces or enhancing the local infrastructure to be "cloud-ready." See also return on investment below.

- Pay per use. The capability to build up cost according to the actual consumption of resources is a relevant feature of cloud systems. Pay per use strongly relates to quality of service support, where specific requirements to be met by the system, hence to be paid for, can be specified. One of the key economic drivers for the current level of interest in Cloud Computing is the structural change in this domain. By moving from the usual capital upfront investment model to an operational expense, Cloud Computing promises to enable small and medium enterprises (SMEs) and entrepreneurs especially to accelerate the development and adoption of innovative solutions.

- Improved time to market is essential in particular for SMEs that want to sell their services quickly and easily with little delays caused by acquiring and setting up the infrastructure, in particular in a scope compatible and competitive with larger industries. Larger enterprises need to be able to publish new capabilities with little overhead to remain competitive. Clouds can support this by providing infrastructures, potentially dedicated to specific use cases that take over essential capabilities to support easy provisioning and thus reduce time to market.

- Return on investment (ROI) is essential for all investors and cannot always be guaranteed; in fact, some cloud systems currently fail in this aspect. Employing a cloud system must ensure that the cost and effort invested in it is outweighed by its benefits of being commercially viable—this may entail direct (e.g., more customers) and indirect (e.g., benefits from advertisements) ROI. Outsourcing resources versus increasing the local infrastructure and employing (private) cloud technologies needs to be outweighed and critical cut-off points identified.

- Turning CAPEX into OPEX is an implicit and much argued characteristic of cloud systems, as the actual cost benefit (cf. ROI) is not always clear. Capital expenditure (CAPEX) is required to build up a local infrastructure, but with outsourcing computational resources to cloud systems on demand and scalable, a company will actually spend operational expenditure (OPEX) for provisioning of its capabilities, as it will acquire and use the resources according to operational need.

- "Going green" is relevant not only to reduce additional costs of energy consumption, but also to reduce the carbon footprint. While carbon emission by individual machines can be quite well estimated, this information is actually taken less into consideration when scaling systems up. Clouds principally allow reducing the consumption of unused resources (downscaling). In addition, up scaling should be carefully balanced not only with cost, but also with carbon emission issues. Note that beyond software stack aspects, plenty of green IT issues are subject to development on the hardware level.

6.2.2.2 Non-Functional Aspects

Unlike economic or technological aspects, non-functional aspects represent the qualities or properties of a system, rather than specific technological requirements. Implicitly, they can be realized in multiple fashions and interpreted in different ways. This typically leads to strong compatibility and interoperability issues between individual providers as they pursue their own approaches to realize their respective requirements, which strongly differ between providers. Non-functional aspects are some of the key reasons why "clouds" differ so strongly in their interpretation.

The most important non-functional aspects are [3]:

- Elasticity: Elasticity is an essential core feature of cloud systems and circumscribes the capability of the underlying infrastructure to adapt to changing, potentially non-functional requirements; for example, amount and size of data supported by an application, number of concurrent users, etc. One can distinguish between horizontal and vertical scalability, whereby horizontal scalability refers to the amount of instances to satisfy changing amount of requests, and vertical scalability refers to the size of the instances themselves and thus is implicit to the amount of resources required to maintain the size. Cloud scalability involves both (rapid) up- and downscaling. Elasticity goes a step further and allows for the dynamic integration and extraction of physical resources to the infrastructure. While from the application perspective, this is identical to scaling, from the middleware management perspective this poses additional requirements, in particular regarding reliability. In general, it is assumed that changes in the resource infrastructure are announced first to the middleware manager, but with large-scale systems, it is vital that such changes can be maintained automatically.

- Reliability: Reliability is essential for all cloud systems—in order to support today's data center-type applications in a cloud, reliability is considered one of the main features to exploit cloud capabilities. Reliability denotes the capability to ensure constant operation of the system without disruption, that is, no loss of data, no code reset during execution, etc. Reliability is typically achieved through redundant resource utilization. Interestingly, many of the reliability aspects have moved from hardware to a software-based solution (redundancy in the file systems vs. RAID controllers, stateless front-end servers vs. UPS, etc.). Notably, there is a strong relationship between availability (see below) and reliability; however, reliability focuses in particular on prevention of loss (i.e., data or execution in progress).

- Quality of service (QoS) support: This relevant capability is essential in many use cases where specific requirements have to be met by the outsourced services or resources. In business cases, basic QoS metrics like response time, throughput, etc. must be guaranteed at least to ensure that the quality guarantees of the cloud user are met. Reliability is a particular QoS aspect that forms a specific quality requirement.

- Agility and adaptability: These essential features of cloud systems strongly relate to the elastic capabilities. They involve on-time reaction to changes in the amount of requests and size of resources, but also adaptation to changes in the environmental conditions that require different types of resources, different quality or different routes, etc. Implicitly, agility and adaptability require resources (or at least their management) to be autonomic and have to enable them to provide self-* capabilities like self detection of failure, self awareness, self activity, self elasticity etc.

- Availability of services and data: This is an essential capability of cloud systems and was actually one of the core aspects that gave rise to clouds in the first instance. It lies in the ability to introduce redundancy for services and data so failures can be masked transparently. Fault tolerance also requires the ability to introduce new redundancy (e.g., previously failed or fresh nodes) in an online manner non-intrusively (without a significant performance penalty). With increasing concurrent access, availability is particularly achieved through replication of data services and distributing them across different resources to achieve load balancing. This can be regarded as the original essence of scalability in cloud systems.

6.2.2.3 Technological Aspects

Technological challenges implicitly arise from non-functional and economical aspects, when trying to achieve them [3]. As opposed to these aspects, technological challenges typically imply a specific realization—even though there may be no standard approach as yet and deviations may hence arise. In addition to these implicit challenges, one can identify additional technological aspects to be addressed by cloud systems, partially as a pre-condition to achieving some of the high-level features, but partially also as they directly relate to specific characteristics of cloud systems.

The main technological challenges that can be identified and that are commonly associated with cloud systems are:

- Virtualization: This is an essential technological characteristic of clouds, which hides the technological complexity from the user and enables enhanced flexibility (through aggregation, routing, and translation). More concretely, virtualization supports the following features:

  - Ease of use: Through hiding the complexity of the infrastructure (including management, configuration, etc.), virtualization can make it easier for the user to develop new applications, as well as reduce the overhead for controlling the system.

  - Infrastructure independency: In principle, virtualization allows for higher interoperability by making the code platform independent.

- Flexibility and adaptability: By exposing a virtual execution environment, the underlying infrastructure can change more flexibly according to different conditions and requirements (assigning more resources, etc.).

- Location independence: Services can be accessed independent of the physical location of the user and the resource.

- Multi-tenancy: This is a highly essential issue in cloud systems, where the location of code and data is principally unknown and the same resource may be assigned to multiple users (potentially at the same time). This affects infrastructure resources as well as data/applications/services that are hosted on shared resources but need to be made available in multiple isolated instances. Classically, all information is maintained in separate databases or tables, yet in more complicated cases information may be concurrently altered, even though maintained for isolated tenants. Multi-tenancy implies many potential issues, ranging from data protection to legislator issues.

- Security, privacy and compliance: These are obviously essential in all systems dealing with potentially sensitive data and code.

- Data management: This is an essential aspect in particular for storage clouds, where data is flexibly distributed across multiple resources. Implicitly, data consistency needs to be maintained over a wide distribution of replicated data sources. At the same time, the system always needs to be aware of the data location (when replicating across data centers) taking latencies and particularly workload into consideration. As size of data may change at any time, data management addresses both horizontal and vertical aspects of scalability. Another crucial aspect of data management is provided consistency guarantees (eventual vs. strong consistency, transactional isolation vs. no isolation, atomic operations over individual data items vs. multiple data times, etc.).

- APIs or programming enhancements: These are essential to exploit the cloud features: common programming models require that the developer takes care of the scalability and autonomic capabilities, while a cloud environment provides the features in a fashion that allows the user to leave such management to the system.

- Metering of any kind of resource and service consumption is essential in order to offer elastic pricing, charging, and billing. It is therefore a pre-condition for the elasticity of clouds.

- Tools are generally necessary to support development, adaptation, and usage of cloud services.

6.2.3 Service Models

Beyond the base definition, NIST goes on to explain that Cloud Computing consists of three service models, which can be offered across three different deployment models. The service models consist of Software as a Service (SaaS), Platform as a Service (PaaS), and Infrastructure as a Service (IaaS).

6.2.3.1 Software as a Service (SaaS)

NIST's definition of SaaS is "The capability provided to the consumer is to use the provider's applications running on a cloud infrastructure. The applications are accessible from various client devices through either a thin client interface, such as a web browser (e.g., web-based email), or a program interface. The consumer does not manage or control the underlying cloud infrastructure including network, servers, operating systems, storage, or even individual application capabilities, with the possible exception of limited user-specific application configuration settings." In other words, SaaS consists of firms offering the capability to use software applications that are housed off the user's premises. Examples are the customer relationship management tools that Salesforce.com provides and the office productivity suites available from Google. These services are a natural outgrowth of the software programs that have been resident on and used in standalone computing environments.

6.2.3.2 Platform as a Service (PaaS)

The capability provided to the consumer is to deploy onto the cloud infrastructure consumer-created or acquired applications created using programming languages, libraries, services, and tools supported by the provider. The consumer does not manage or control the underlying cloud infrastructure including network, servers, operating systems, or storage, but has control over the deployed applications and possibly configuration settings for the application-hosting environment. PaaS services allow users to develop their own web-based applications or to customize existing applications using one or more programming languages and development tools. Examples include Amazon's Elastic Compute Cloud and Google's App engine. These services are natural extensions of individual computer platforms like Linux and Windows.

6.2.3.3 Infrastructure as a Service (IaaS)

The capability provided to the consumer is to provision processing, storage, networks, and other fundamental computing resources where the consumer is able to deploy and run arbitrary software, which can include operating systems and applications. The consumer does not manage or control the underlying cloud infrastructure but has control over operating systems, storage, and deployed applications; and possibly limited control of select networking components (e.g., host firewalls). IaaS services allow customers to access the equipment and hardware needed to perform computing operations, including storage, processing, and networking components.

6.2.4 Deployment Models

6.2.4.1 Private Cloud

The cloud infrastructure is provisioned for exclusive use by a single organization comprising multiple consumers (e.g., business units). It may be owned, managed, and operated by the organization, a third party, or a combination of both, and it may exist on or off premises.

Private clouds are typically owned by the respective enterprise or leased. Functionalities are not directly exposed to the customer, although in some cases services with cloud-enhanced features may be offered.

This deployment model is typically the first choice for companies during their first steps into the cloud. Because such clouds are exclusively used by a single organization, private clouds can be considered most secure. On the other hand, the possibility for productivity gains compared to multi-tenant cloud deployment models is minimal.

6.2.4.2 Community Cloud

The cloud infrastructure is provisioned for exclusive use by a specific community of consumers from organizations that have shared concerns (e.g., mission, security requirements, policy, and compliance considerations). It may be owned, managed, and operated by one or more of the organizations in the community, a third party, or some combination of these, and it may exist on or off premises.

Because multiple parties are sharing the same infrastructure, this option can be considered as less secure compared to private clouds. This deployment model offers higher economies of scale (compared to private clouds).

6.2.4.3 Public Cloud

The cloud infrastructure is provisioned for open use by the public. It may be owned, managed, and operated by a business, academic or government organization, or a combination of these. It exists on the premises of the cloud provider.

This deployment model can be considered as less secure than private and community clouds, but it has potential for large economies of scale.

6.2.4.4 Hybrid Cloud

The cloud infrastructure is a composition of two or more distinct cloud infrastructures (private, community, or public) that remain unique entities, but are bound together by standardized or proprietary technology that enables data and application portability (e.g., cloud bursting for load balancing between clouds).

While the choice of deployment model has implications for the security and privacy of a system [4], the deployment model itself does not dictate the level of security and privacy of specific cloud offerings. That level depends mainly on assurances, such as the soundness of the security and privacy policies, the robustness of the security and privacy controls, and the extent of visibility into performance and management details of the cloud environment, which are furnished by the cloud provider or independently attained by the organization (e.g., via independent vulnerability testing or auditing of operations).

6.2.5 Summary

It is difficult to come up with a precise definition of Cloud Computing. In general terms, the idea is that you do not maintain a part of the IT architecture (meaning hardware, data storage, and software) by yourself. This infrastructure is not located at the site where it is used and it can be rented as a service from different cloud providers. Software and data as well are not locally located or in the corporation's data center. Everything is in the "cloud." Access to those resources takes place via a network, for example, Internet or a corporate intranet (private cloud).

6.3 VARIOUS USE CASES FOR CLOUD COMPUTING

Cloud Computing can be used in various areas. Here, we summarize several basic use cases to highlight how important Cloud Computing can be for companies.

6.3.1 e-Business

Scalability plays an important role in e-business. Nowadays most customers buy almost everything online. Especially during high shopping seasons or particular periods (weekends, evening) it is important that web shop homepages are available to any potential customers. Otherwise, this would mean losses. To ensure this availability, new virtual servers can be made available when needed (e.g., every evening, every weekend, and during high shopping seasons as well).

Depending on user needs or priorities, rules can also be used to specify which applications require more resources than others do. For example, if a company uses the cloud both for e-business and for research, users can assign e-business applications a higher priority because such applications generate revenues.

6.3.2 Pilot Phase

A new functionality, software, or even operating system can be tested during a so-called pilot phase. For this pilot phase, resources are needed as well. The request can be made online through a simple web interface. After specification of the start and end dates of the pilot phase, the request goes to a cloud resource administrator who approves or rejects the request. In case of approval, the cloud provisions, servers, and resources are available within minutes.

6.3.3 Virtual Worlds

Virtual worlds are another usage scenario for the cloud. An example of a virtual world is the so-called "Massively Multiplayer Online Games" (MMPOG), which is a multiplayer video game capable of supporting hundreds or thousands of players simultaneously. By necessity, the games are played on the Internet, and feature at least one persistent world [8]. The more users log in, the more computing power is needed.

Due to the massive amount of concurrent MMPOG players, there is a need for large numbers of reliable infrastructure (servers). The added value of Cloud Computing in this case can be:

- monitoring the infrastructure's current usage level

- applying load balancing rules between servers whenever it is needed

- individual innovations

The pilot phase as described above is not a concept developed and owned by companies and businesses. More individuals are coming up with innovations and may need servers from a cloud to work on their innovations.

Besides providing a computing platform or substitute for in-house applications, public cloud services such as the following can also focus on augmenting security in other computing environments [4]:

- Data Center Oriented—Cloud services can be used to improve the security of data centers. Information about online activities collected from many participants from different organizations can allow for better threat monitoring. For example, electronic mail can be redirected to a cloud provider via mail exchange (MX) records, and examined and analyzed collectively with similar transactions from other data centers to discover widespread spam, phishing, and malware campaigns, and to carry out remedial action (e.g., quarantining suspect messages and content) more comprehensively than a single organization would be able to do.

- Cloud Oriented—Cloud services can also be used to improve the security of other cloud environments. Cloud-based identity management services exist, which can be used to augment or replace an organization's directory service for identification and authentication of users to a cloud.

With any technology area, the functionality afforded can be turned toward improper or illicit activities. Cloud computing is no exception. A couple of noteworthy instances have already occurred that give a sense of what might be expected in the future:

- Botnets—A botnet is a collection of Internet-connected computers whose security defenses have been breached and control ceded to a malicious party. Each such compromised device, known as a "bot," is created when a computer is penetrated by software from a malware distribution, otherwise known as malicious software [5].

Botnets are mostly assembled and controlled by hackers and many characteristics of Cloud Computing (cost reduction, dynamic provisioning, redundancy, security) apply to them. Botnets could be used to launch a denial of service attack against the infrastructure of a cloud provider.

- Mechanism Cracking—WiFi Protected Access (WPA) Cracker, a cloud service ostensibly for penetration testers, is an example of harnessing cloud resources on demand to break a cryptographic cipher and determine the encrypted password used to protect a wireless network. In effect, because cryptography is used widely in authentication, data confidentiality, integrity, and other security mechanisms, these become less effective with the availability of cryptographic key cracking cloud services. Both cloud-based and traditional types of systems are possible targets.

- Completely Automated Public Turing test to tell Computers and Humans Apart (CAPTCHA) cracking is another area where cloud services could be applied to bypass verification meant to thwart abusive use of Internet services by automated software. CAPTCHA involves the solution of a simple test by a user before accessing a service, as a means of thwarting unwanted automated access.

6.4 BENEFITS

Thanks to Cloud Computing, companies are faced with completely different business models that rely on robust web applications and networks as well. This new operation model is not always the best fit for all customers but it comes with key benefits, which we will discuss in this section.

- The "pay for what you use" strategy makes IT costs transparent. Customers avoid unnecessary expenses, for example, in hardware, software, and tools. Monitoring and controlling resource usage provides transparency for both the provider and the consumer. Customers can easily see if resources are unused and release them in order to keep the bill low.

- Better customer satisfaction triggered by an increased reliability of cloud services. Because cloud providers offer a high quality of service, one can think that the risk of common threats like network outages is reduced. Users can also immediately react in case of emergency.

- Cloud providers' expertise. The provider can specialize in certain applications and cloud services. By doing this they gain expertise and knowledge, which can lead to more efficient management of several tasks like maintenance, upgrade, backup, etc.

- Better flexibility due to the possibility of on-the-fly requests of resources. IT departments that previously had to struggle for additional hardware and software during high fluctuation periods can now request and release capacity as needed.

- Infrastructure monitoring allows organizations to make ad-hoc adjustments in order to increase or decrease capacity. In this way, customers do not need to pay for unused capacity and providers can assign the freed resources to other customers.

- Added value in scalability. It is now possible to request large storage space or important processing capacity in order to pursue certain goals during a limited time, for example. Some universities will not have to invest in an expensive infrastructure that will only be needed once. This allows departments with restricted finances the ability to run research projects without the burden of high costs.

- It is no longer mandatory to have one's own IT infrastructure in order to offer specific IT services. This is a great opportunity for small companies. They can have access to state-of-the-art resources at low prices and do not need to "waste" financial resources for IT infrastructure. In addition, countries with limited access to real technological infrastructure can use the Internet (if existing) and access virtual resources via the cloud.

- Open standards. Several cloud computing experts advocate for more standards that are open, open software, and data interoperability in the cloud. This vision of Cloud Computing would make it more comfortable for customers to switch between different cloud providers without having to adapt their data from the previous provider to the new provider. In addition, accessing cloud services from different platforms (desktop, laptop, tablet, etc.) would be much easier.

- Resource pooling. Cloud providers can maximize the load on the shared resources according to consumer demand by reassigning unused resources to other consumers. Because it is easy to request resources and these resources can be quickly assigned, projects can be started quickly as well.

- Energy efficiency. Resource pooling makes consumers share resources. Customers no longer need to run their own data center/IT infrastructure, which implies reduced energy consumption on the consumer side and maximized load on the provider side.

Besides the benefits listed previously, there is also a high potential for security and privacy improvement in Cloud Computing. Computer and network security issues observed in traditional computing also apply in cloud systems. These issues are the need to maintain the integrity of data, the need to ensure that systems are available, and the need to provide data confidentiality. It is understandable that companies would think twice before giving up direct control of IT infrastructure to a cloud provider.

Some cloud services (e.g., cloud backup services) require a combination of resources and uniform security management practices. This makes new security improvement possible for clouds. For example, clouds can detect and predict new threats and provide quick solutions as well.

Smaller organizations with limited numbers of IT administrators and security personnel seem to be the biggest beneficiaries and they can gain the economies of scale available to larger organizations with sizeable data centers by transitioning to a public cloud.

Opportunities for improved security also benefit privacy. That is, effective privacy can exist only upon a sound foundation of information security. However, privacy, just as security, has broad organizational, operational, and technical implications. While some aspects of privacy are closely related to the confidentiality, integrity, and availability objectives of security, other aspects are not.

Potential areas of improvement where organizations may derive security and privacy benefits from transitioning to a public Cloud-Computing environment include the following [4]:

- Staff Specialization—Cloud providers, just as other organizations with large-scale computing facilities, have an opportunity for staff to specialize in security, privacy, and other areas of high interest and concern to the organization. Increases in the scale of computing induce specialization, which in turn allows security staff to shed other duties and concentrate exclusively on security and privacy issues. Through increased specialization, there is an opportunity for staff members to gain in-depth experience and training, take remedial actions, and make improvements to security and privacy more readily than otherwise would be possible with a more diverse set of duties.

- Platform Strength—The structure of Cloud Computing platforms is typically more uniform than that of most traditional computing centers. Greater uniformity and homogeneity facilitate platform hardening and enable better automation of security management activities like configuration control, vulnerability testing, security audits, and security patching of platform components. Information assurance and security response activities also profit from a uniform, homogeneous cloud infrastructure, as do system management activities, such as fault management, load balancing, and system maintenance. Similarly, infrastructure homogeneity benefits management controls employed to protect privacy. On the other hand, homogeneity means that a single flaw will be manifested throughout the cloud, potentially affecting all tenants and services. Many Cloud Computing environments meet standards for operational compliance and certification in areas such as healthcare (e.g., Health Insurance Portability and Accountability Act [HIPAA]), finance (e.g., Payment Card Industry Data Security Standard [PCI DSS]), security (e.g., ISO 27001, Information Security Management Systems Requirements), and audit (e.g., Standards for Attestation Engagements [SSAE] No. 16), and may attain formal certification or attestation from an independent third party to impart a level of assurance with regard to some recognized and generally accepted criteria.

- Resource Availability—The scalability of Cloud Computing facilities allows for greater availability. Redundancy and disaster recovery capabilities are built into Cloud Computing environments and on-demand resource capacity can be used for better resilience when faced with increased service demands or distributed denial of service attacks, and for quicker recovery from serious incidents. When an incident occurs, an opportunity also exists to contain attacks and capture event information more readily, with greater detail and less impact on production. In some cases, however, such resilience and capacity can have a downside. Access to vast amounts of inexpensive storage may also engender more information to be collected than needed or information to be retained longer than necessary.

- Backup and Recovery—The backup and recovery policies and procedures of a cloud provider may be superior to those of the organization and may be more robust. Data maintained within a cloud can be easily available, faster to restore, and more reliable under many circumstances than that which is maintained in a traditional data center. Cloud data also have the advantage of meeting offsite backup storage and geographical compliance requirements. However, network performance over the Internet and the amount of data involved are limiting factors that can affect restoration.

- Mobile Endpoints—The architecture of a cloud solution extends to the client at the service endpoint that is used to access hosted applications. Cloud clients can be general-purpose web browsers or more special-purpose applications. Since the main computational resources needed by cloud-based applications are typically held by the cloud provider, clients can generally be lightweight computationally and easily supported on laptops, notebooks, and netbooks, as well as embedded devices such as smart phones and tablets, benefiting the productivity of an increasingly mobile workforce.

- Data Concentration—Data maintained and processed in a public cloud may present less of a risk to an organization with a mobile workforce than having that data dispersed on portable computers, embedded devices, or removable media out in the field, where theft and loss routinely occur. That is not to say, however, that no risk exists when data is concentrated. Many organizations have made the transition to support access to organizational data from mobile devices to improve workflow management and gain other operational efficiencies and productivity benefits. Carefully constructed applications can restrict access and services to only the data and tasks that correspond strictly with the responsibilities a user needs to accomplish, limiting data exposure in the event of a device compromise.

6.5 RISKS

With Cloud Computing, several significant concerns arise. Delicate topics like data integrity, security, privacy, intellectual property management, and audit trails are relevant and must be handled accordingly. These concerns are complex because they involve different parties like institutional policies and state regulations. Because cloud services consumers lose control over IT infrastructure, the success of Cloud Computing initiatives relies on trust between the customer and the service provider. Nevertheless, answering the following questions can help consumers to make their choice.

- What possibilities are offered by the providers in order to react to any incident affecting security or privacy?
- How secure is the provider's IT infrastructure?
- How secure are the provider's applications and processes?
- Which privacy policy is applied for hosted data and applications?
- How is personal information collected and used?

6.5.1 Security and Privacy

Security and privacy become an issue when agencies and organizations start considering the transition of applications and data to public Cloud Computing environments. Therefore, we will discuss the problem faced in public clouds regarding security and privacy.

6.5.1.1 Key Privacy and Security Issues

Although the emergence of Cloud Computing is a recent development [4], insights into critical aspects of security can be gleaned from reported experiences of early adopters and from researchers analyzing and experimenting with available cloud provider platforms and associated technologies. The following sections highlight privacy and security-related issues that are believed to have long-term significance for public Cloud Computing and, in many cases, for other Cloud Computing service models. Where possible, examples of previously exhibited or identified problems are provided to illustrate an issue. These examples are not exhaustive and may cover only one aspect of a more general issue. For many of the issues, the specific problems discussed have been resolved. Nevertheless, broader issues persist in most cases, which are potentially re-expressed in other ways among the various service models. Security and privacy considerations that stem from IT outsourcing also exist; they are covered in the next section and complement the following material.

Because Cloud Computing has grown out of an amalgamation of technologies, including service-oriented architecture, virtualization, Web 2.0, and utility computing, many of the privacy and security issues involved can be viewed as known problems cast in a new setting. The importance of their combined effect in this setting, however, should not be discounted. Public Cloud Computing represents a thought-provoking paradigm shift from conventional norms to an open deperimeterized organizational infrastructure—at the extreme, displacing applications from one organization's infrastructure to the infrastructure of another organization, where the applications of potential adversaries may also operate.

- Governance—Governance implies control and oversight by the organization over policies, procedures, and standards for application development and IT service acquisition, as well as the design, implementation, testing, use, and monitoring of deployed or engaged services. With the wide availability of Cloud Computing services, lack of organizational controls over employees engaging in such services arbitrarily can be a source of problems. While Cloud Computing simplifies platform acquisition, it does not alleviate the need for governance; on the contrary—it amplifies that need.

- Compliance—Compliance refers to an organization's responsibility to operate in agreement with established laws, regulations, standards, and specifications. Various types of security and privacy laws and regulations exist within different countries at national, state, and local levels, making compliance a potentially complicated issue for Cloud Computing. For example, at the end of 2010, the National Conference of State Legislatures reported that 46 states have enacted legislation governing disclosure of security breaches of personal information, and that at least 29 states have enacted laws governing the disposal of personal data held by businesses or government.

- Trust—Under the Cloud Computing paradigm, an organization relinquishes direct control over many aspects of security and privacy, and in so doing, confers a high level of trust onto the cloud provider. At the same time, federal agencies have a responsibility to protect information and information systems, commensurate with the risk and magnitude of harm that may result from unauthorized access, use, disclosure, disruption, modification, or destruction. This is the case regardless of whether the information is collected or maintained by or on behalf of an agency, or whether the information systems are used or operated by an agency or by a contractor of an agency or other organization on behalf of an agency.

6.5.2 Security and Privacy Downside

Besides its many potential benefits for security and privacy [4], public Cloud Computing also brings with it potential areas of concern, when compared with computing environments found in traditional data centers. Some of the more fundamental concerns include the following:

- System Complexity—A public Cloud Computing environment is extremely complex compared with that of a traditional data center. Many components make up a public cloud, resulting in a large attack surface. Besides components for general computing, such as deployed applications, virtual machine monitors, guest virtual machines, data storage, and supporting middleware, there are also components that the management backplane comprises, such as those for self-service, resource metering, quota management, data replication and recovery, service level monitoring, workload management, and cloud bursting. Cloud services themselves may also be realized through nesting and layering with services from other cloud providers. Components change over time as upgrades and feature improvements occur, confounding matters further.

- Security depends not only on the correctness and effectiveness of many components, but also on the interactions among them. Challenges exist in understanding and securing application-programming interfaces that are often proprietary to a cloud provider. The number of possible interactions between components increases as the square of the number of components, which pushes the level of complexity upward. Decreases in security also heighten privacy risks related to the loss or unauthorized access, destruction, use, modification, or disclosure of personal data.

- Shared Multi-Tenant Environment—Public cloud services offered by providers have a serious underlying complication—client organizations typically share components and resources with other consumers that are unknown to them. Rather than using physical separation of resources as a mode of control, Cloud Computing places greater dependence on logical separation at multiple layers of the application stack. While not unique to Cloud Computing, logical separation is a non-trivial problem that is exacerbated by the scale of Cloud Computing. An attacker could pose as a consumer to exploit vulnerabilities from within the cloud environment, overcome the separation mechanisms, and gain unauthorized access.

- Threats to network and computing infrastructures continue to increase each year and become more sophisticated. Having to share an infrastructure with unknown outside parties can be a major drawback for some applications and require a high level of assurance pertaining to the strength of the security mechanisms used for logical separation.

- Internet-facing Services—Public cloud services are delivered over the Internet, exposing the administrative interfaces used to self-service and manage an account, as well as non-administrative interfaces used to access deployed services. Applications and data that were previously accessed from the confines of an organization's intranet, but moved to a public cloud, must now face increased risk from network threats that were previously defended against at the perimeter of the organization's intranet and from new threats that target the exposed interfaces. The performance and quality of services delivered over the Internet may also be an issue. The effect is somewhat analogous to the inclusion of wireless access points into an organization's intranet at the onset of that technology, necessitating additional safeguards for secure use.

- Relying on remote administrative access as the means for the organization to manage assets that are held within the cloud also increases risk, compared with a traditional data center, where administrative access to platforms can be restricted to direct or internal connections. Similarly, remote administrative access of the cloud infrastructure, if done by the cloud provider, is also a concern. When taken together with the previous two items, a highly complex, multi-tenanted computing environment, whose services are Internet-based and available to the public, arguably affords a potentially attractive attack surface that must be carefully safeguarded.

- Loss of Control—While security and privacy concerns in Cloud Computing services are similar to those of traditional non-cloud services, they are amplified by external control over organizational assets and the potential for mismanagement of those assets. Transitioning to a public cloud requires a transfer of responsibility and control to the cloud provider over information as well as system components that were previously under the organization's direct control. The transition is usually accompanied by a lack of a direct point of contact with the management of operations and influence over decisions made about the computing environment. This situation makes the organization dependent on the cooperation of the cloud provider to carry out activities that span the responsibilities of both parties, such as continuous monitoring and incident response. Compliance with data protection laws and regulations is another important area of joint responsibility that requires coordination with and the cooperation of the cloud providers.

- Loss of control over both the physical and logical aspects of the system and data diminishes the organization's ability to maintain situational awareness, weigh alternatives, set priorities, and effect changes in security and privacy that are in the best interest of the organization. Legal protections for privacy may also be affected when information is stored with a third-party service provider. Under such conditions, maintaining accountability can be more challenging, offsetting some of the potential benefits discussed previously.

6.5.3 Audit Trails

Audit trails [7] maintain a record of system activity both by system and application processes and by user activity of systems and applications. In conjunction with appropriate tools and procedures, audit trails can assist in detecting security violations, performance problems, and flaws in applications.

Audit trails may be used as either a support for regular system operations or a kind of insurance policy or as both of these. As insurance, audit trails are maintained but are not used unless needed, such as after a system outage. As a support for operations, audit trails are used to help system administrators ensure that the system or resources have not been compromised by hackers, insiders, or technical problems.

6.5.3.1 Benefits and Objectives

Audit trails can provide a means to help accomplish several security-related objectives, including individual accountability, reconstruction of events, intrusion detection, and problem analysis.

6.5.3.1.1 Individual Accountability   Audit trails are a technical mechanism that helps managers maintain individual accountability. By advising users that they are personally accountable for their actions, which are tracked by an audit trail that logs user activities, managers can help promote proper user behavior. Users are less likely to attempt to circumvent security policy if they know that their actions will be recorded in an audit log.

For example, audit trails can be used in concert with access controls to identify and provide information about users suspected of improper modification of data (e.g., introducing errors into a database). An audit trail may record "before" and "after" versions of records. (Depending upon the size of the file and the capabilities of the audit logging tools, this may be very resource-intensive.) Comparisons can then be made between the actual changes made to records and what was expected. This can help management determine if errors were made by the user, by the system or application software, or by some other source.

Audit trails work in concert with logical access controls, which restrict use of system resources. Granting users access to particular resources usually means that they need that access to accomplish their job. Authorized access, of course, can be misused, which is where audit trail analysis is useful. While users cannot be prevented from using resources to which they have legitimate access authorization, audit trail analysis is used to examine their actions. For example, consider a personnel office in which users have access to personnel records for which they are responsible. Audit trails can reveal that an individual is printing far more records than the average user, which could be an indication that personal data are being sold. Another example may be an engineer who is using a computer to design a new product. Audit trail analysis could reveal that the engineer used an outgoing modem extensively the week before quitting. This could be used to investigate whether proprietary data files were sent to an unauthorized party.

6.5.3.1.2 Reconstruction of Events   Audit trails can also be used to reconstruct events after a problem has occurred. Damage can be more easily accessed by reviewing audit trails of system activity to pinpoint how, when, and why normal operations ceased. Audit trail analysis can often distinguish between operator-induced errors (during which the system may have performed exactly as instructed) or system-created errors (e.g., arising from a poorly tested piece of replacement code). If, for example, a system fails or the integrity of a file (either program or data) is questioned, an analysis of the audit trail can reconstruct the series of steps taken by the system, the users, and the application. Knowledge of the conditions that existed at the time of a system crash, for example, can be useful in avoiding future outages. Additionally, if a technical problem occurs (e.g., the corruption of a data file), audit trails can aid in the recovery process (e.g., by using the record of changes made to reconstruct the file).

6.5.3.1.3 Intrusion Detection   Intrusion detection refers to the process of identifying attempts to penetrate a system and gain unauthorized access. If audit trails have been designed and implemented to record appropriate information, they can assist in intrusion detection. Although normally thought of as a real-time effort, intrusions can be detected in real time by examining audit records as they are created (or through the use of other kinds of warning flags/notices) or after the fact (e.g., by examining audit records in a batch process).

Real-time intrusion detection is primarily aimed at outsiders attempting to gain unauthorized access to the system. It may also be used to detect changes in the system's performance indicative of, for example, a virus or worm attack (forms of malicious code). There may be difficulties in implementing real-time auditing, including unacceptable system performance.

After-the-fact identification may indicate that unauthorized access was attempted (or was successful). Attention can then be given to damage assessment or reviewing controls that were attacked.

6.5.3.1.4 Problem Analysis   Audit trails may also be used as online tools to help identify problems other than intrusions as they occur. This is often referred to as real-time auditing or monitoring. If a system or application is deemed critical to an organization's business or mission, real-time auditing may be implemented to monitor the status of these processes (although, as noted previously, there can be difficulties with real-time analysis). An analysis of the audit trails may be able to verify that the system operated normally (i.e., that an error may have resulted from operator error, as opposed to a system-originated error). Such use of audit trails may be complemented by system performance logs. For example, a significant increase in the use of system resources (e.g., disk file space or outgoing modem use) could indicate a security problem.

6.5.3.2 Audit Trails and Logs

A system can maintain several different audit trails concurrently. There are typically two kinds of audit records: (1) an event-oriented log and (2) a record of every keystroke, often called keystroke monitoring. Event-based logs usually contain records describing system events, application events, or user events.

An audit trail should include sufficient information to establish what events occurred and who (or what) caused them. In general, an event record should specify when the event occurred, the user ID associated with the event, the program or command used to initiate the event, and the result. Date and time can help determine if the user was a masquerader or the actual person specified.

6.5.3.2.1 Keystroke Monitoring   Keystroke monitoring is the process used to view or record both the keystrokes entered by a computer user and the computer's response during an interactive session. Keystroke monitoring is usually considered a special case of audit trails. Examples of keystroke monitoring include viewing characters as they are typed by users, reading users' electronic mail, and viewing other recorded information typed by users.

Some forms of routine system maintenance may record user keystrokes. This could constitute keystroke monitoring if the keystrokes are preserved along with the user identification so that an administrator could determine the keystrokes entered by specific users. Keystroke monitoring is conducted in an effort to protect systems and data from intruders, who access the systems without authority or in excess of their assigned authority. Monitoring keystrokes typed by intruders can help administrators assess and repair damage caused by intruders.

6.5.3.2.2 Audit Events   System audit records are generally used to monitor and fine-tune system performance. Application audit trails may be used to discern flaws in applications or violations of security policy committed within an application. User audits records are generally used to hold individuals accountable for their actions. An analysis of user audit records may expose a variety of security violations, which might range from simple browsing to attempts to plant Trojan horses or gain unauthorized privileges.

The system itself enforces certain aspects of policy (particularly system-specific policy) such as access to files and access to the system itself. Monitoring the alteration of systems configuration files that implement the policy is important. If special accesses (e.g., security administrator access) have to be used to alter configuration files, the system should generate audit records whenever these accesses are used.

Sometimes a finer level of detail than system audit trails is required. Application audit trails can provide this greater level of recorded detail. If an application is critical, it can be desirable to record not only aspects that invoked the application, but also certain details specific to each use. For example, consider an e-mail application. It may be desirable to record who sent the mail as well as the receiver of the mail and the length of the message. Another example would be that of a database application. It may be useful to record who accessed a particular database as well as the individual rows or columns of a table that were read, changed, or deleted, instead of just recording the execution of the database program.

A user audit trail monitors and logs user activity in a system or application by recording events initiated by the user (e.g., access of a file, record, or field, use of a modem).

Flexibility is a critical feature of audit trails. Ideally, from a security point of view, a system administrator would have the ability to monitor all system and user activity, but could choose to log only certain functions at the system level, and within certain applications. The decision of how much to log and how much to review should be a function of application/data sensitivity and should be decided by each functional manager/ application owner with guidance from the system administrator and the computer security manager/officer, weighing the costs and benefits of the logging. Audit logging can have privacy implications; users should be aware of applicable privacy laws, regulations, and policies that may apply in such situations.

- System-level Audit Trails—If a system-level audit capability exists, the audit trail should capture, at a minimum, any attempt to log on (successful or unsuccessful), the log-on ID, date and time of each log-on attempt, date and time of each log-off, the devices used, and the functions performed once logged on (e.g., the applications that the user tried, successfully or unsuccessfully to invoke). System-level logging also typically includes information that is not specifically security-related, such as system operations, cost-accounting charges, and network performance.

- Application-Level Audit Trails—System-level audit trails may not be able to track and log events within applications, or may not be able to provide the level of detail needed by application or data owners, the system administrator, or the computer security manager. In general, application-level audit trails monitor and log user activities, including data files opened and closed, specific actions, such as reading, editing, and deleting records or fields, and printing reports. Some applications may be sensitive enough from data availability, confidentiality, or integrity perspective that a "before" and "after" picture of each modified record (or the data elements changed within a record) is captured by the audit trail.

- User Audit Trails—User audit trails can usually log:

  - all commands directly initiated by the user,

  - all identification and authentication attempts, and

  - files and resources accessed.

It is most useful if options and parameters are also recorded from commands. It is much more useful to know that a user tried to delete a log file (e.g., to hide unauthorized actions) than to know the user merely issued the delete command, possibly for a personal data file.

6.5.3.3 Implementation Issues

Audit trail data requires protection because the data should be available for use when needed and is not useful if it is not accurate. In addition, the best-planned and implemented audit trail is of limited value without timely review of the logged data. Audit trails may be reviewed periodically, as needed (often triggered by occurrence of a security event), automatically in real-time, or in some combination of these. System managers and administrators, with guidance from computer security personnel, should determine how long audit trail data would be maintained—either on the system or in archive files.

Next are examples of implementation issues that may be addressed when using audit trails.

6.5.3.3.1 Protecting Audit Trail Data   Access to online audit logs should be strictly controlled. Computer security managers and system administrators or managers should have access for review purposes. However, security or administration personnel who maintain logical access functions may have no need for access to audit logs.

It is particularly important to ensure the integrity of audit trail data against modification. One way to do this is to use digital signatures. Another way is to use write-once devices. The audit trail files need to be protected because, for example, intruders may try to "cover their tracks" by modifying audit trail records. Audit trail records should be protected by strong access control mechanisms in order to prevent unauthorized access. The integrity of audit trail information may be particularly important when legal issues arise, such as when audit trails are used as legal evidence (e.g., this may require daily printing and signing of the logs). Questions of such legal issues should be directed to the cognizant legal counsel.

The confidentiality of audit trail information may also be protected, for example, if the audit trail is recording information about users that may be disclosure-sensitive such as transaction data containing personal information (e.g., "before" and "after" records of modification to income tax data). Strong access controls and encryption can be particularly effective in preserving confidentiality.

6.5.3.3.2 Review of Audit Trails   Audit trails can be used to review what occurred after an event, for periodic reviews, and for real-time analysis. Reviewers should know what to look for to be effective in spotting unusual activity. They need to understand what normal activity looks like. Audit trail review can be easier if the audit trail function can be queried by user ID, terminal ID, application name, date and time, or some other set of parameters to run reports of selected information.

6.5.3.3.2.1 *Audit Trail Review after an Event*   Following a known system or application software problem, a known violation of existing requirements by a user, or some unexplained system or user problem, the appropriate system-level or application-level administrator should review the audit trails. Review by the application/data owner would normally involve a separate report, based upon audit trail data, to determine if their resources are being misused.

6.5.3.3.2.2 Periodic Review of Audit Trail Data   Application owners, data owners, system administrators, data processing function managers, and computer security managers should determine how much review of audit trail records is necessary, based on the importance of identifying unauthorized activities. This determination should have a direct correlation to the frequency of periodic reviews of audit trail data.

6.5.3.3.2.3 Real-Time Audit Analysis   Traditionally, audit trails are analyzed in a batch mode at regular intervals (e.g., daily). Audit records are archived during that interval for later analysis. Audit analysis tools can also be used in a real-time or near real-time fashion. Such intrusion detection tools are based on audit reduction, attack signature, and variance techniques. Manual review of audit records in real-time is almost never feasible on large multiuser systems due to the volume of records generated. However, it is possible to view all records associated with a particular user or application, and view them in real time. This is similar to keystroke monitoring, although it may be legally restricted.

6.5.3.3.3   Tools for Audit Trail Analysis   Many types of tools have been developed to help reduce the amount of information contained in audit records, as well as to distill useful information from the raw data. Especially on larger systems, audit trail software can create very large files, which can be extremely difficult to analyze manually. The use of automated tools is likely to be different between an unused audit trail data and a robust program.

Audit reduction tools are preprocessors designed to reduce the volume of audit records to facilitate manual review. Before a security review, these tools can remove many audit records known to have little security significance. This alone may cut the number of records in the audit trail in half. These tools generally remove records generated by specified classes of events, such as records generated from night backups.

Trends/variance-detection tools look for anomalies in user or system behavior. It is possible to construct more sophisticated processors that monitor usage trends and detect major variations. For example, if a user typically logs in at 9 a.m., but appears at 4:30 a.m. one morning, this may indicate a security problem that may need to be investigated.

Attack signature-detection tools look for an attack signature, which is a specific sequence of events indicative of an unauthorized access attempt. A simple example would be repeated failed login attempts.

6.5.3.4 Cost Considerations

Audit trails involve many costs. First, some system overhead is incurred recording the audit trail. Additional system overhead will be incurred storing and processing the records. The more detailed the records, the more overhead is required. Another cost involves human and machine time required for the analysis. This can be minimized using tools to perform most of the analysis. Many simple analyzers can be constructed quickly (and economically) from system utilities, but they are limited to audit reduction and identifying only particularly sensitive events. More complex tools that identify trends or sequences of events are slowly becoming available as off-the-shelf software. If complex tools are not available for a system, development may be prohibitively expensive. Some intrusion detection systems, for example, have taken years to develop.

The final cost of audit trails is the cost of investigating anomalous events. If the system is identifying too many events as suspicious, administrators may spend undue time in the reconstruction of events and questioning of personnel.

6.6 IS CLOUD COMPUTING GREEN?

The answer to this question depends on what is meant by "green."

The color green is associated with paper money. Therefore, "green cloud" can be used to describe the capability of cloud systems to be cost-efficient. In this context, it is correct to say that Cloud Computing is green thanks to the following key features:

- Cost reduction—Compared to traditional IT, an important cost reduction is observed in Cloud Computing from the customer point of view. No additional investment is needed to buy hardware. Maintenance cost does not apply anymore because there is nothing to maintain. Energy consumption is reduced, which leads to a considerable improvement in the yearly electricity bill.

- Pay per use—Thanks to this feature, it is possible to pay only for the required services. Existing monitoring features give a manager the ability to oversee the utilized services. In this way, it is also possible to cancel obsolete services and lower the related costs.

The other meaning of green is "environmentally friendly." In this context, I am not sure if it is correct to say that Cloud Computing is green because the energy consumption is just outsourced from the consumer's former data center to the cloud provider. Cloud Computing can make it possible to maximize the usage of resources thanks to resource sharing. However, if the resources are to be used simultaneously, the cloud service provider has to buy more resources in order to satisfy the need, leading to more energy consumption. Since the energy consumption is centralized at the cloud provider's data center, one can say that it is better to have many parties share the same infrastructure without seeing the corresponding energy consumption multiplied by 100. However, the more people who share resources, the more additional resources have to be afforded in order to satisfy the need and keep the service quality and the more energy is consumed. In my point of view, the best way to be environmentally friendly would be not to consume energy at all. But that is another topic.

REFERENCES

[1] Mell, P. and Grance, T. 2011. The NIST Definition of Cloud Computing.
[2] Software & Information Industry Association. 2001. Software As A Service: Strategic Backgrounder.
[3] CORDIS. 1994–2012. The future of Cloud Computing, http://cordis.europa.eu/, ©European Union.
[4] Jansen, W. and Grance, T. 2011. NIST Special Publication 800-144 Guidelines on Security and Privacy in Public Cloud Computing.
[5] Wikipedia. 2012. http://en.wikipedia.org/wiki/Botnet.
[6] NIST Special Publication 800-53A Revision 1. 2010. Information Security.
[7] NIST Special Publication 800-12. 1995. An Introduction to Computer Security: The NIST Handbook.
[8] Wikipedia. http://en.wikipedia.org/wiki/Massively_multiplayer_online_game. November 2012.

Index